Pelican Books
Metals in the Service of Man

Dr Arthur Street and Professor William Alexander
both graduated in the early nineteen-thirties from
Birmingham University, where they studied
metallurgy and conducted researches. Until his
retirement in 1975, Dr Street was Chairman of a
well-known diecasting company. He wrote one of the
first books about diecasting and has recently produced
a modern one. He has lectured on diecasting in many
parts of the world. In 1975 he became the first European
to win the Doehler Award, given annually by the
American Die Casting Institute. His principal hobby
is Grand Opera.

For many years Professor Alexander was connected
with production and metallurgical research in one of
the largest non-ferrous metal organizations in
Britain, where at various times he worked on copper,
aluminium, lead, zinc, iron, uranium, thorium,
niobium and beryllium, and some of their alloys. From
1967 until his retirement in 1976, he was Professor of
Metallurgy at the University of Aston in Birmingham,
where he concentrated on developing course work on
industrial metallurgy and engineering materials,
especially on the application, conservation, energy
utilization and economic aspects.

Arthur Street
William Alexander

Metals in the
Service of Man

SIXTH EDITION

Penguin Books

Penguin Books Ltd, Harmondsworth,
Middlesex, England
Penguin Books, 625 Madison Avenue,
New York, New York 10022, U.S.A.
Penguin Books Australia Ltd, Ringwood,
Victoria, Australia
Penguin Books Canada Ltd, 2801 John Street,
Markham, Ontario, Canada L3R 1B4
Penguin Books (N.Z.) Ltd, 182–190 Wairau Road,
Auckland 10, New Zealand

First published 1944
Reprinted 1945, 1946
Second edition 1951
Third edition 1954
Reprinted 1956, 1958, 1960
Fourth edition 1962
Reprinted with revisions 1964
Reprinted 1965
Reprinted with revisions 1968
Reprinted 1969
Fifth edition 1972
Reprinted 1973
Sixth edition 1976
Reprinted 1977

Made and printed in Great Britain
by Richard Clay (The Chaucer Press) Ltd, Bungay, Suffolk
Set in Monotype Plantin

To Brenda and Kath, with many thanks for their help, encouragement and endurance since the first edition of this book was written

CONTENTS

PREFACE

The science of metals is a specialized one and tends to be shut away from the general reader in rather grim-looking text books and papers. This is a pity, because metals retain the attraction and wonder they had for us in our youth and they are of great importance to mankind. We hope this book will interest all those who handle metals in their leisure, pleasure, or daily work, whether this is in the home, industry, school or university. We have tried to avoid the 'sensational discovery' style and have not hesitated to include discussions of some complex subjects. At the same time we have attempted to present our material in a readable form; where technical terms have been introduced they have been defined in the glossary.

The chapter about metals in nuclear energy was introduced in 1964, and we thank those who helped us to present this subject in what we hope is a clear manner. In 1971 the currency of Britain went decimal; engineering, commerce and education were beginning to think metric and to use the international SI units. We therefore decided to change all measurements to metric, with sufficient information about the conversions to the new units to help those who find it difficult to visualize the strength of metals in terms of newtons per square millimetre.

During the last thirty years, much useful information for revised editions has been provided by members of the Institute of Metals and the Iron and Steel Institute (which were merged to form the Metals Society in January 1974). Readers from all parts of the world have written to us with helpful suggestions – and sometimes with corrections. We again thank all these and we remain sincerely grateful to those friends who, in the early nineteen-forties, helped and encouraged us to embark on the first edition of this book which, we understand, has persuaded many young people that metallurgy as a career is interesting and rewarding.

A.C.S., W.O.A.

'DRAMATIS PERSONAE'

Principals

IRON	The most important metal
ALUMINIUM	The light metal
COPPER	The conductivity metal
ZINC	The galvanizing metal
LEAD	The plumbers' metal
TIN	The metal that tins the can
NICKEL	The versatile metal

Supporting Characters

MAGNESIUM, BERYLLIUM	The ultra-light metals
TITANIUM	The strong middleweight
CHROMIUM	The stainless metal
TUNGSTEN	The lamp filament metal
GOLD, SILVER, PLATINUM	The precious trio
IRIDIUM, PALLADIUM, RHODIUM, RUTHENIUM	The valuable quartet
GERMANIUM	A transistor element
TANTALUM	The capacitor metal
MANGANESE, VANADIUM	The scavenging metals
COBALT, MOLYBDENUM	Some other metals that give new properties to steel
CADMIUM	The weather resister
OSMIUM	The heaviest metal
LITHIUM	The lightest metal
BARIUM, CAESIUM, CALCIUM, POTASSIUM, SODIUM	Some reactive metals
MERCURY	The liquid metal
ANTIMONY, ARSENIC, BISMUTH, BORON, TELLURIUM	Semi-metals
SILICON, SELENIUM	Semi-conductor elements

NIOBIUM, ZIRCONIUM	Some new arrivals
GALLIUM, HAFNIUM, INDIUM, RHENIUM, RUBIDIUM, SCANDIUM	Some rare metals
CERIUM, DYSPROSIUM, ERBIUM, EUROPIUM, GADOLINIUM, HOLMIUM, LANTHANUM, LUTETIUM, NEODYMIUM, PRASEODYMIUM, PROMETHIUM, SAMARIUM, TERBIUM, THULIUM, YTTERBIUM, YTTRIUM,	The 'rare earth' metals
PLUTONIUM, RADIUM, THORIUM, URANIUM	The radioactive metals
CARBON, OXYGEN	Two non-metals, vital to life and useful to metallurgy

LIST OF PLATES

A NOTE ON METRICATION

For precise measurements, the following typical conversions to metric or SI units have been used.

1 pound	equals	0·453 kilogram
1 inch	equals	25·4 millimetres
1 gallon	equals	4·5 litres
1 ton force per sq. inch	equals	15·44 newtons per sq. millimetre

The derivation of newtons, an important unit of force in connection with tensile strength of metals, is discussed on page 116. In order to help get acquainted with these new units, we have shown the strength of metals in tons per sq. inch for comparison, in several parts of the book.

Tons or Tonnes?

The 1,000 kilogram tonne is equal to 2,204 pounds and is therefore about 1·6 per cent lighter than an Imperial ton. Most industrial concerns are already purchasing metal by the tonne. We guess that the word 'tonne' will not be accepted at all readily in Britain, owing to its 'olde worlde' flavour, though the measure of 1,000 kilograms will soon be universally employed. We have, therefore, compromised. Where a precise weight is indicated, particularly for the cost of metals, we have used tonnes. Where an approximate weight is mentioned, as for example the annual world production of a metal several years ago, we use the word 'tons'.

Distances

For long distances, no great precision was needed, so we used the motorist's conversion of eight kilometres to five miles. For precise measurements we convert inches to millimetres with a factor of 25·4, or to centimetres with a factor of 2·54. Intermediate measurements have been converted to metres.

Temperatures

All previous editions of this book have given temperatures in degrees Centigrade. We believe that British people, having laboriously learned not to use Fahrenheit, will be reluctant to use the words 'Celsius' or 'Kelvin', especially as the units of Kelvin and Celsius temperature intervals are identical with those of Centigrade. A temperature expressed in degrees Centigrade or Celsius is equal to the temperature expressed in Kelvin less 273·15 degrees (Kelvin's zero is Absolute zero).

Pressures

1 ton per sq. in. is converted to 15·44 newtons per sq. mm. However those engineers who deal with pressures – as for example boiler engineers – are being rather slow to accept SI units so we apologize if we have anticipated their acceptance on some occasions where we refer to pressures.

General

So far as possible we have tried to avoid pedantry. For example, where approximate measurements have been given we have converted approximately. Thus where in a previous edition we described some long Roman nails as 16 inches long, we have described them as 400 millimetres long in this edition. As most plant engineers still use horse power, we have retained that. In referring to telescope mirrors in the chapter on aluminium we have retained 'inches' because that still seems the current practice in astronomy, though the space programme uses metric measurements.

I
METALS AND CIVILIZATION

Although metal-working was not the first craft known to mankind, our present material civilization is derived from our knowledge and use of metals. From the time of Tubal Cain* onwards, the early users of metals would probably have described them as bright substances which were hard but could be hammered into various useful shapes. All the metals which they knew were heavy – iron, for instance, being nearly three times as heavy as granite. If the metal-worker was also a hunter or warrior he might speak with satisfaction of the sharpness of metal weapons and of their endurance through many combats. Craftsmen in wood or in stone would praise the new metal cutting-tools which made their jobs so much easier. Thus, even in the very early days, there grew an appreciation of the value of metals.

Man learned to make fires hot enough to melt metals in earthenware containers, which the Romans called 'crucibuli' and which we now call 'crucibles'. He discovered that the molten metal could be poured into the cavity made by placing together two halves of a hollowed-out clay or stone mould: the metal filled the cavity and, when solid, it was found to have taken the shape of the cavity. Archaeologists have discovered ancient bronze swords and arrow-heads, made by casting in this manner.

The art of blending metals was gradually developed and it became known that an 'alloy' formed in this way was sometimes stronger, harder, and tougher than the metals of which it was composed. Probably the first alloy to be made was a bronze, consisting of copper, with about one part in ten of tin. The primitive metallurgists discovered that if a greater proportion of

* Genesis iv, 22.

tin were used the alloy was harder, while less tin gave a softer alloy, so that for different purposes bronzes with varying tin contents were deliberately produced. By the time the Romans came to Britain, they were using iron and bronze for weapons, tools, and farming implements; copper for vessels and ornaments; lead for water pipes, baths, and even coffins; tin, gold, and silver for ornaments; and silver, brass, and bronze for coinage.

Gold and, to a lesser extent, silver, were called 'noble metals' because they could be exposed to the atmosphere for a long time without tarnishing and because they could be melted repeatedly without much loss in weight. These characteristics led to their being used for jewellery and eventually coinage. The possession of noble metals consequently became a measure of wealth so that gold and silver were coveted for their monetary as distinct from their utilitarian value. All the other metals then known, such as tin, lead, copper, and iron, were by contrast considered 'base metals'.

Owing to the general similarities between all metals it was perhaps natural to imagine that one metal could be changed into another, and it seemed particularly desirable that base metals should be transmuted into noble ones – gold for preference. Beginning in the Middle East in the early part of the Christian era, and thriving in that part of the world and in Europe till the end of the seventeenth century, the art of alchemy became associated with the search for the elusive Philosophers' Stone, which was supposed to be capable of turning base metals into gold or silver.

As recently as the end of the eighteenth century a writer defined alchemy as 'a science and art of making a fermentative powder which transmutes imperfect metals into gold and which serves as a universal remedy for all the ills of man, animals, and plants'. The philosopher Francis Bacon was nearer to the truth when he commented: 'Alchemy may be compared to the man who told his sons that he had left them gold buried somewhere in his vineyard; where they by digging found no gold, but by turning up the mould about the roots of the vines, procured a plentiful vintage. So the search and endeavours to make gold have brought many useful inventions and instructive experiments to light.'

While the alchemists and their wealthy, and often exasperated, patrons were trying all kinds of experiments to produce the Philosophers' Stone, the metal-workers, though using fewer incantations, were developing almost as wonderful a process – the changing of dull, earthy minerals or 'ores' into metals by smelting them with charcoal in a fire or furnace. They learned how to recognize those metallic ores which could be smelted profitably and how to transform them into metal. It is not surprising that their efforts were sometimes unsuccessful. Even today, producers of metal encounter difficulties because of the presence of impurities in the ore, which upset smelting operations or have a harmful influence on the resulting metal. In those early days such happenings were freely attributed to the Evil Eye, or to the attention of hobgoblins, or Old Nick himself. The name of nickel was derived from this, while the name for cobalt originated from the German elves called *kobolds*.

Between the Middle Ages and the beginning of the industrial era, the main progress in the art of making metals was the building of larger and more efficient furnaces to produce metal in greater quantity. In 1740, when Dr Johnson used to drink tea with Sir Joshua Reynolds and Mr Garrick, Britain, which was the world's greatest producer of metals, made less than twenty thousand tons of iron per year.* (We make that amount in about eight hours in this country now.) A hundred years later, when the penny post had just been introduced, Britain made one and a quarter million tons of iron annually, and by the end of the nineteenth century the figure had risen to about nine million tons per year.

About twenty years after the beginning of Victoria's reign, the manufacture of steel was becoming a major industry; in 1856 Henry Bessemer made public his process for converting large quantities of pig iron into steel. The rapid development of the use of iron and steel throughout the civilized world was out-

* Anyone who is interested in the metallurgical beginnings of the Industrial Revolution should visit the Ironbridge Gorge and the splendid open-air museum at Blists Hill, between Birmingham and Shrewsbury.

standing, though other metals, such as copper, tin, lead and zinc, were also being produced in increasing quantities. Metals in general, and steel in particular, came to be used for making bridges, railway lines, ships, guns, implements of all kinds, and, towards the end of the nineteenth century, those noisy 'horseless carriages' which chugged their way along the roads at a rattling speed sometimes exceeding five miles per hour.

Queen Victoria's reign therefore extended over a period during which important developments occurred in the history of metals; the *art* of metal-working was slowly growing into the *science* of metallurgy. In 1861 Professor Henry C. Sorby of Sheffield initiated the systematic examination of metals, using the microscope, and thus laid the foundations of that branch of metallurgy called metallography. Properties of metals and alloys – their melting points, strength, hardness, and electrical properties – were studied and correlated. A new, unusually light metal, aluminium, was discovered; nickel and then other metals were alloyed with steel to give improved properties.

In the twentieth century the expansion of the use of metals has been so rapid that considerably more metal has been extracted during the last seventy years than during the ages from the beginning of man's history till A.D. 1900. *Table 1* shows a comparison of the world average production of some well-known metals, first for two fifty-year periods – 1858–1907 and 1908–1957 – and then during the period 1958–1970.

How many people realize that our material civilization and every amenity of life depend on the work of the metallurgist and on his ability to produce the right metal for each particular purpose? In the morning we are awakened by an alarm clock, the working components of which are metallic. We press an electric switch and current passes along a copper wire to light a lamp with a tungsten filament. We wash in water which has come through copper, lead or stainless-steel pipes, shave with a stainless-steel razor-blade, whilst anticipating breakfast that is to be cooked on a pressed-steel stove; meanwhile tea is being made in a metal tea-pot, with water heated in an aluminium kettle. Then, whilst eating our bacon and egg, with the aid of a stainless-steel knife and fork, we read a newspaper that has been printed with a

Table 1. World Production of the Common Metals

Metal	Period 1858–1907 Average tons per annum	Period 1908–1957 Average tons per annum	Period 1958–1970 Average tons per annum
Copper	260,000	1,700,000	6,000,000
Lead	500,000	1,400,000	2,500,000
Tin	60,000	140,000	150,000
Zinc	300,000	1,400,000	3,300,000
Aluminium	2,000	760,000	8,000,000
Magnesium	NIL	40,000	155,000
Nickel	3,500	30,000	470,000
Pig iron	35,000,000	40,000,000	330,000,000
Steel	11,000,000	120,000,000	475,000,000

lead alloy type-metal. Having finished breakfast, we rush to catch a bus or train which is almost entirely made of metal. We pay a 10p coin made of an alloy of copper and nickel, and receive change in copper–zinc–tin alloy pence. This is only the beginning of a day during which hundreds of metal objects may be used, whether we operate a lathe, supervise a computer or drive a tractor.

Our material civilization depends on the efficient harnessing of power, but the control of this power is made possible by the use of metal and alloys. Without metals no railway, aeroplane, motor car, electric motor or interplanetary space vehicle could operate. Three of the latest and greatest achievements of men are structures made of at least 95 per cent metal, namely the Apollo Rockets, new suspension bridges such as those over the River Severn and the Bosphorus, and the Concorde Supersonic Aircraft (Plate 5a).

The variety of metals which is now available has certainly benefited mankind, but it has made discrimination necessary to get the best use from each metal or alloy. For example Concorde was designed to withstand the arduous conditions of aerodynamic heating coupled with the long life of 45,000 flying hours required of the first supersonic aircraft for airline use. 71 per cent of the structure is made of a special aluminium alloy, RR.58, which can withstand these conditions. 16 per cent of the weight of Concorde is of high strength steel, for the undercarriage. Titanium alloys, amounting to 4 per cent, are used in the engine nacelle; this has to withstand a temperature of about 400°C; such a temperature is too high for aluminium alloys, but not so high that heat-resisting steels, or nickel alloys, would have been required. The remaining 9 per cent comprises nickel alloys, plastics, glass and other materials.

During the last twenty-five years many metals have emerged from being laboratory curiosities to a position of some importance: tantalum, titanium, niobium, zirconium; these and many others are extending the repertoire of the metallurgist. The requirements of nuclear energy, jet aircraft and space ships have provided incentives to overcome the immense problems of winning these metals from their ores and of fabricating them. Sometimes completely new methods have been developed to shape metals which, a generation ago, were thought quite intractable.

The rigid classification of the elements into metals (iron, copper) and non-metals (carbon, sulphur) is no longer completely valid. Elements such as boron behave sometimes as non-metals, sometimes as metals; they are known as semi-metals or metalloids. The semi-conductors, including germanium and selenium, have special electrical properties that have led to their use in transistors and optical mechanisms.

While development of the new metals is proceeding, fresh endeavours are continually being made to widen the scope of the well-known materials. New ways of producing steel, new alloys, greater purity of metals, often revealing unexpected properties, improved methods of welding and cutting metals, are just a few of the developments which have taken place in the last quarter century. In these and other ways the designer and engineer are being offered a wide range of metallic materials. Using the specialized knowledge and advice of the metallurgist they are able to select the most suitable metal for the job and the most suitable treatment for that metal, bearing in mind its mechanical and physical properties, its ease of fabrication, availability and cost.

Modern civilized mankind is so familiar with metals that the difficulty of defining a metal may be overlooked. Indeed, different branches of science have different definitions. A chemist might say 'Metallic elements are those which possess alkaline hydroxides'. A physicist might define a metal as 'An element with a high electrical conductivity which decreases slightly with increase of temperature'. Someone who works with metals would remind us that they possess lustre when polished and are good conductors of electricity and heat; he would add that most of them are denser, stronger, and more malleable than the other, non-metallic, elements.

Engineers will tell you that in general metals are the only materials worth considering for constructing any mechanism or structure which must have a high efficiency: for example, a motor car. This is because many metals and alloys have great strength and capability of withstanding limited over-loading without catastrophic failure; in other words, they are tough.

We, the authors, having handled a variety of metals and alloys for over forty years, can tell you of our abiding life interest. We can reassure those following that there is still much scope for craft and ingenuity in metallurgy, despite the fact that great strides have been made in it as a science during the past seventy years, and also that real fundamental understanding and valid explanations of many of the properties of metals which are used by man still call for intensive scientific work.

2

HOW WE GET
OUR METALS

When Commander Robert E. Peary was exploring Greenland in 1894 an Eskimo took him to a place near Cape York where he found three metallic meteorite masses. The Eskimos called them Saviksue, or 'the great irons', each having a name suggested by its shape. Peary removed 'The Woman' and 'The Dog' to his ship in 1895, and brought them to the American Museum of Natural History. The largest meteorite, 'The Tent', was much more difficult to remove, but in 1897 Peary and his men transferred the great mass to his ship. His four-year-old daughter, Marie Ahnighito Peary, threw a bottle of wine on the meteorite, which was then re-named Ahnighito. In 1904 it was hauled slowly along Broadway and 77th Street to the entrance of the American Museum. When the Hayden Planetarium was opened in 1935, Ahnighito made its last journey, halfway around Manhattan Square to the Planetarium first floor, where it has remained on exhibit ever since. The meteorite weighs about thirty tons and is composed of an alloy of iron, nickel and cobalt.

Although meteorites of such size do not often fall to the earth from outer space, smaller ones have been found quite frequently, usually consisting of iron, containing about 8 per cent of nickel, with a small amount of cobalt. No doubt primitive man, whose local culture was thus by accident raised from the level of the stone age to that of the iron age, thought metallic meteorites were valuable gifts from the gods. Nowadays, however, meteorites are hardly regarded as a useful source of iron. For one thing the delivery service is erratic and the unheralded arrival of a meteorite in one's back garden would be more embarrassing than profitable.

Copper has also been discovered occasionally in the metallic form. The largest solid mass of pure copper was found in 1856 in what was called the Minesota Mine, in the upper Michigan peninsula. It weighed about 500 tons and was so large that it had to be cut into a number of pieces before being hoisted to the surface. Such a find is, however, of little significance as a source of copper, of which over seven million tons are made annually. Most metals, with the exception of some precious ones, usually appear in nature as minerals or ores, where they exist in chemical combination with other elements. These ores do not resemble the metals which can be extracted from them; one would hardly imagine that strong and bright metals could be produced from such uninteresting-looking earthy substances. Yet it is easy to realize that rust is a chemical compound of iron, while verdigris is a compound of copper; these are the kinds of metal-bearing materials which mineral deposits contain. The ores are treated by fire, chemical or electrical processes, to convert them into metals, these operations being known as 'smelting'. Before that can be done, there is a chain of processes; suitable ores must be discovered, they must be mined or quarried, concentrated and the ore deposits smelted.

METALLIC ORES

It is believed that the central core of the earth, about 6,500 km in diameter, consists of a molten alloy of iron and nickel, at a temperature of 3,700°C. Perhaps one day this may provide a source of molten metal, like a Texan oil gusher, but at the moment there would be problems. Mining metallurgists win their ores from the outer crust of the earth.

In 1924 two American scientists, F. W. Clarke and H. S. Washington, attempted to estimate the proportion of the chemical elements contained in the outer ten miles of the earth's crust. To find the approximate composition of the 17,000,000,000,000,000,000 tons of materials comprising this outer skin of the earth, a series of averages was taken, based on 5,159 chemical analyses of rocks all over the world. Such an investigation could not provide exact results, but Clarke and

Washington's figures give an approximate idea of the relative amounts of the elements contained near the surface of the earth. *Fig. 1* illustrates this analysis of the earth's crust.

Oxygen and silicon, which are present in granites, sandstones, and the majority of rocks, make up the major part of the composition of the earth's crust, the elements, of course, being in chemical combination with others. Nearly half the weight is oxygen, while silicon accounts for more than a quarter. Aluminium, iron and magnesium also exist in large quantities. Some comparatively unfamiliar metals occur in considerable amounts; for example, titanium makes up one part in 160, which means there are over 100,000,000,000,000,000 tons of titanium contained in the earth's crust. On the other hand copper, tin, nickel, zinc, lead, mercury, silver, antimony and gold, which are essential to our needs and civilization, are among some of the rarest elements in the earth's crust. Thus copper is present to the extent of only one part in ten thousand and lead only one part in fifty thousand.

The availability of metals does not depend on the amount present in the earth, but on the ease with which their ores can be obtained and smelted. An ore deposit near Chicago or Newcastle upon Tyne is likely to be more useful than similar ones in the tropical forests of the Amazon or at the bottom of the ocean under the Mindanao Trench. The largest iron ore reserves, in Brazil, are estimated to be 50,000 million tons – about one third of the world's surface stock – but, because of their inaccessibility, these have begun to be exploited only recently.

A customer of the local inn would doubtless agree that half a pint in one glass is preferable to sixty teaspoonfuls in sixty glasses, and an analogy can be made with regard to the use of metallic ores. A single rich, extensive ore deposit of a comparatively scarce metal can be exploited more readily than many small pockets spread all over the globe. Although ores of copper, lead, zinc, and nickel form only a small proportion of the earth's crust, they are often found in locally rich deposits which can be mined in sufficient quantities to supply man's present needs.

Although one twelfth of the earth's crust by weight consists of

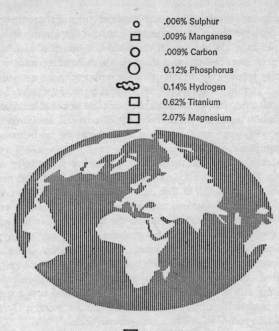

.006% Sulphur
.009% Manganese
.009% Carbon
0.12% Phosphorus
0.14% Hydrogen
0.62% Titanium
2.07% Magnesium

2.58% Potassium

2.83% Sodium

3.64% Calcium

5.06% Iron

8.07% Aluminium

27.61% Silicon

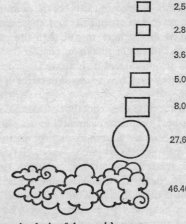

46.46% Oxygen

Fig. 1. Analysis of the earth's crust

aluminium, many of the compounds of this metal are unsuitable to be used as ores. Ordinary garden-clay comprises about 25 per cent aluminium, in chemical combination with silicon, oxygen, iron, calcium and magnesium. A hundred barrow loads of clay would contain enough aluminium to make a small aeroplane, but the refining processes would be too expensive. So long as lavish deposits of bauxite, the principal ore of aluminium, are available, such as those recently discovered in Jamaica and Australia, we need not concern ourselves with extracting aluminium from clays.

Metallic ores are distributed somewhat irregularly in the earth. Some parts, such as Australia, Mexico, and land near the Great Lakes in the U.S.A., are well supplied with a variety of useful deposits. Others, such as Arabia and Palestine, are less well endowed, though the salt in the Dead Sea may be regarded as a potential source of magnesium. There are some places where Nature has been lavish, as, for example, the Atach Mountain in the Urals, which has been described as a vast lump of iron ore and near which the U.S.S.R. built the great iron-making city of Magnitogorsk. Although Britain is not specially rich in metallic ores, save those of iron, one county, Cornwall, contains some ore of every well-known metal.

Figs. 2, 3, and 4 attempt to give 'score cards' of the countries where ore deposits of various metals are mined; first, those major metals which are described in chapters 10–16; then the most important among what we call minor metals in chapter 17. In most cases the five biggest ore-producers are indicated in order of magnitude of output. The tonnages are expressed in terms of the metal contents of the ores.

The majority of useful metallic ores contain the metal combined with oxygen, sulphur, or other elements, as shown in *Table 2*. Pure minerals are rarely found in nature; they are generally contaminated with gravel, limestone, sand, clay, and stones. This unwanted material is termed 'gangue' (pronounced to rhyme with 'hang'). Although some ores are comparatively rich, others contain large amounts of gangue; for example, copper ores may have only 1 per cent of metal, nickel ores about 3 per cent, and zinc ores about 10 per cent.

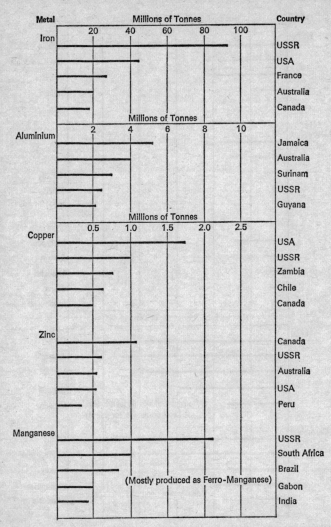

Fig. 2. The first five countries which produce the most ore for some important metals

(*The tonnages represent metal content of the ores*)

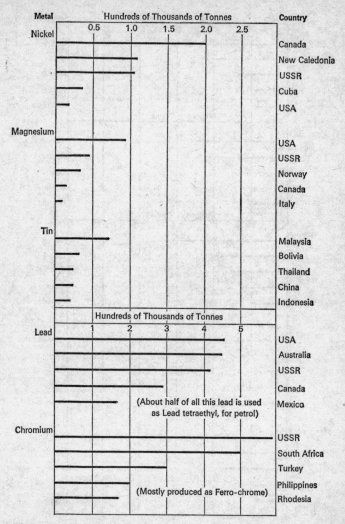

Fig. 3. The first five countries which produce the most ore for some other important metals

(*The tonnages represent metal content of the ores*)

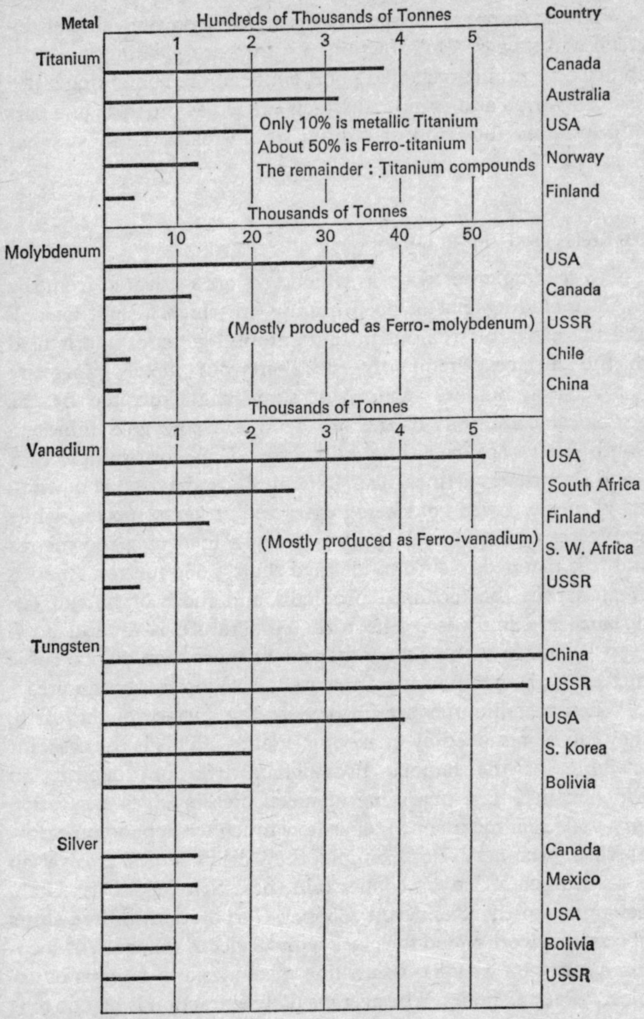

Fig. 4. The first five countries which produce the most ore for some minor metals

(*The tonnages represent metal content of the ores*

Whether an ore is classed as rich depends on the value of the metal and its ease of extraction. An iron ore yielding less than 20 per cent iron is reckoned poor, while an ore containing 2 per cent tin is rich and a mineral which carried as much as one part in a hundred thousand of radium would make the most sober Board of Directors rejoice.

OBTAINING THE ORE

There are numerous ways in which ores are recovered from the earth, including that of deep mining, in which a shaft is sunk and the ore is blasted out in underground galleries and hauled to the surface. Frequently the shafts for metal mines are not vertical, but are inclined at some angle dictated by the geological conditions of the ore deposit. Some gold mines in South Africa are over three kilometres deep, for example East Rand Proprietary Mines, near Johannesburg, have gone down to 3,430 metres, with gold being extracted at 3,200 metres, while at Western Deep Levels gold is being mined at 3,260 metres and it is planned to sink an inclined shaft 3,660 metres. At such great depths the technical problems and those of human endurance are immense. The rock temperature is around 49°C (120°F) and in each of the larger mines more than 70,000 cubic metres of air, much of it refrigerated, is required per minute.

When metallic ores are on or near the surface of the earth, they can be removed by quarrying (Plate 3). This is the case, for example, at the famous Frodingham iron ore deposits in Lincolnshire. The quarrying of metal ores involves excavation on a scale that makes one realize how much we depend on earth-moving machinery. For example, the world's largest excavation is the Bingham Canyon copper mine near Salt Lake City, Utah, belonging to the Kennecott Copper Corporation. Three shifts of 900 men work round the clock with 38 electric shovels, 62 locomotives, 1,268 wagons, 18 drilling machines and 28 tons of explosives per 24 hours. The average daily extraction is 96,000 tons of one per cent ore and 225,000 tons of overburden. Plate 2 shows an extraordinary aerial view of this mine; the concentric rings are the terraces of the mine, which are excavated by power shovels.

Table 2. Some Minerals from which Metals are Obtained

Metal	Name of mineral contained in the ore	Metallic compound contained in mineral	
		Chemical name	Chemical symbols
Aluminium	Bauxite	Hydrated aluminium oxide	$Al_2O_3.3H_2O$
Copper	Copper pyrites	Copper–iron sulphide	$CuFeS_2$
Iron	Haematite	Iron oxide	Fe_2O_3
Lead	Galena	Lead sulphide	PbS
Magnesium	Magnesite	Magnesium carbonate	$MgCO_3$
	Dolomite	Magnesium–calcium carbonate	$MgCa(CO_3)_2$
Mercury	Cinnabar	Mercury sulphide	HgS
Nickel	Pentlandite	Nickel–iron sulphide	$NiS(FeS)_2$
Silver	Argentite	Silver sulphide	Ag_2S
Tin	Cassiterite	Tin oxide	SnO_2
Zinc	Zinc blende	Zinc sulphide	ZnS
	Calamine	Zinc carbonate	$ZnCO_3$

In other instances the ore may be in the form of a waterborne deposit, and methods based on the familiar 'panning' principle of the old-time gold miner are employed. A more up-to-date form of panning is that used in the recovery of tin ores in Malaya, where huge dredges dig up the tin-bearing gravel from lake bottoms and pass the wet gravel over long troughs fitted with shallow projections which retain the dense tin ore.

CONCENTRATING THE ORE

Iron ores are comparatively rich, with only 10 to 30 per cent unwanted material – earth, sand, clay and rocks. These contain silica which is almost impossible to melt and which tends to combine with metallic compounds when heated with them. It can be removed during smelting by mixing limestone with the iron ore; in the great heat of the blast furnace, lime combines with the silica, forming a molten slag which can be separated.

Magnesium and aluminium are made by electrical processes, which will not operate unless a very pure metallic compound is separated first. A chemical method of concentrating aluminium ores is described in chapter 4. Ores of copper, zinc, tin and lead contain only a few per cent of metal, so the enormous amount of unwanted material must be removed before smelting. Some weak ores are concentrated by passing them, with a stream of water, over reciprocating metal tables. This causes the light earthy material to move in one direction and the heavier metallic compounds to be delivered to another outlet.

In contrast with this, however, the widely used flotation process is one which usually causes the dense metal compounds to float in a bath of frothed liquid while the unwanted minerals sink. A story, which is probably apocryphal, is told about the discovery of the flotation process. In a certain leadmining village, a woman was washing her husband's working clothes at a tub in her backyard. The washing water was full of suds and the clothes were impregnated with particles of galena, the lead ore which he mined. The peculiar thing about this wash-tub was that the soapsud bubbles, instead of appearing as a white foam, were distinctly dark. Some observant fellow noticed that the surface of the suds was covered with tiny particles of galena. This was surprising, for the lead ore, being very dense, would be expected to fall to the bottom of the tub and not be floating on the surface of the water. The anonymous observer saw in this wash-tub the germ of a commercial process for separating finely crushed lead ore from the gangue associated with it. Americans who tell this story maintain that it occurred in an American mining village; British narrators insist that it was in Derbyshire.

The flotation process has enabled many low-grade ores to be utilized, and it will undoubtedly be employed to an increasing extent as we are forced to use ores of diminishing metal content. The finely crushed ore is agitated in water containing one or more chemical 'frothing' reagents. The earthy matter is usually 'wetted' and sinks, but the metal-containing particles rise in the froth, whence they are skimmed off and dried.

LEACHING THE ORES

Alluvial gold was first discovered about 1000 B.C. Sands containing particles of gold were washed with sheepskins and the metal was retained in the hairs, the lighter sand being washed away. It is believed that when Jason and his Argonauts went seeking the Golden Fleece, their voyage round the Black Sea was in search of gold. In those days the Colchian peoples at the eastern end of the Black Sea collected gold dust, on the fleece of sheep, as it was washed down the river Phasis in the land of Colchis. This ancient process has been superseded by treating gold ores with potassium cyanide. The development of a method of producing this chemical cheaply and of utilizing it in the extraction of gold was a momentous discovery which greatly influenced the development of gold mines and of international trade in various parts of the world.

After crushing to a fine powder, the ore is agitated with a very dilute solution of cyanide in large tanks, each holding about a thousand tons of crushed ore and water. Once the gold has been dissolved by the cyanide, the solution is filtered and the gold precipitated from the solution. Plate 1 shows an aerial photograph of West Driefontein, the world's top gold-producing mine.

FROM ORE TO METAL

As was seen from *Table 2*, several metallic ores contain the metal combined with oxygen; for example the high-grade iron ores contain iron oxide. This metal is cheap because its ores contain a large proportion of iron oxide which can be smelted on a big scale in blast furnaces, with the expenditure of comparatively little fuel, time or labour per ton of iron. On the other hand, although the aluminium ore, bauxite, contains aluminium oxide, it cannot be changed into metal by smelting in a blast furnace. A large amount of energy is needed to separate the aluminium from the oxygen with which it is chemically combined.

Zinc and lead occur in the form of their sulphides. Such ores are first concentrated, then heated in furnaces with access to air, thus converting the sulphide into oxide, which is then reduced to metal by smelting with coke or other form of carbon. Some ores are still more complicated and need extra processes. Details of the smelting of the two best-known metals will be described, first iron and then aluminium. The production of zinc is mentioned on page 215 and of nickel on page 226.

3
MAKING IRON

When an ore containing iron oxide is heated with carbon, metallic iron is produced. The chemical processes taking place in the blast furnace are many and complex but, bringing them to the simplest form:

$$\text{OXIDE OF} + \text{CARBON} \rightarrow \text{IRON} + \text{OXIDE OF}$$
$$\text{IRON} \qquad\qquad\qquad\qquad \text{CARBON}$$
$$\text{(ore)} \qquad \text{(coke)} \qquad \text{(metal)} \quad \text{(gas)}$$

This chemical change is the fundamental process on which iron smelting is based, though other associated reactions take place at the same time. In addition secondary chemical changes convert the earthy non-metallic materials, contained in the ore, into slag.

In early times, iron ore was heated in a charcoal fire (doubtless by chance at first); when the fire died down, a spongy mass of iron remained which could be hammered into shape and used for tools and weapons. Our metallurgical forefathers found that when a high wind was blowing, their fires burned faster and hotter and the iron was produced more rapidly, so eventually bellows were used to increase the supply of air.

Considering that Japan's iron production is now the most efficient in the world, it is interesting to recall that, only a few years before 1914, iron was being made by heating iron ore mixed with charcoal in a V-shaped trough cut in clay, with holes near the bottom for the introduction of a blast of air. After the iron ore and charcoal had been heated together for about three days, the whole contraption was pulled apart and the solid pieces of iron removed. Using such primitive methods, the metal was not melted.

In Britain during the fifteenth century, furnaces of sufficient size were made, and a forced blast of air used, to obtain a temperature high enough to melt the iron, which could then be poured liquid from the furnace. At first it must have been surprising that this 'cast iron' or 'pig iron' was brittle, whereas the forged iron, which had not been melted, could be shaped by hammering. Nevertheless, the cast iron had the advantage that it could be poured into moulds and thus formed into the required shape. It is now understood that when the iron is molten in a blast furnace it absorbs three to five per cent of carbon and other impurities as well. The spongy but forgeable iron did not absorb so much carbon since the temperature attained in the simple furnace was not high enough.

At first charcoal, obtained from wood, was used in smelting iron. One of these furnaces, at Backbarrow, near Windermere, continued to use charcoal until 1925. The furnace, built in 1711, was 6 metres high and $2\frac{1}{2}$ metres square on the outside; it could produce 6 tons of iron per day. Typically, it was built into the side of a hill so that the workers could tip their baskets of ore and fuel into the furnace from a platform on the hillside. The work was done by seven men, aided by their wives and families. The foreman's cottage was adjacent to the furnace, so he could get up during the night, run the liquid metal from the furnace, and return to bed. However, the job had some disadvantages, including the noise and vibration of the waterwheel driven from the River Leven (for the air blast and trip hammers), and the tramps who came to sleep close to the warmth of the furnace.

Early in the twentieth century, the square stone stack, which had been rebuilt in 1770 to contain a circular hearth holding a 6-ton cast, was still being fuelled by charcoal produced in the local woods by about one hundred colliers, but production ceased during the 1914–18 war, when the ironmaster died unexpectedly. Later, attempts were made to restart, using charcoal from the Forest of Dean, but the material proved unsuitable. In 1926 the furnace was converted to coke blast and continued in operation until the early nineteen-sixties. Today the works, not far from the Dolly-Blue factory, lie derelict. However, of the seven charcoal furnaces that worked in south Lakeland, Duddon Furnace

(1736–1867) is being rebuilt by the Cumbria County Council for preservation. Also what used to be called a 'finery-chafery forge', at Stony Hazel, about four miles north of Backbarrow, is being reconstructed. These two sites will eventually represent the most complete examples of the charcoal iron furnace industry in Britain.

In the seventeenth century Dud Dudley attempted, unsuccessfully, to smelt iron with coal. In 1708, Abraham Darby, a young man who was apprenticed in Birmingham and then managed a brass foundry in Bristol, was put in charge of the iron works at Coalbrookdale. A year later he succeeded in smelting iron with coke, instead of charcoal or coal. This achievement provided the starting point for the manufacture of iron rails, the iron bridge over the Severn, iron boats, iron aqueducts, iron buildings and eventually the world-wide industry which now makes over 600 million tonnes of iron per annum.

IRON ORES

Iron forms several oxides, which are available as iron ores. Haematite contains Fe_2O_3 which, if pure, would give 70 per cent iron. A variety of this compound is known as brown haematite which is the same iron oxide loosely combined with molecules of water. Magnetite is a natural magnetic iron ore, the oxide being Fe_3O_4. There are other iron ores formed of iron carbonate.

The world's annual production of iron is more than 600 million tons, so about 1,000 million tons of iron ore is required per annum. The iron ore reserves that are economically available amount to about 275,000 million tons – enough, say, for the next 200 years. There are over 550,000 million tons which are not at present exploitable. However, it will be remembered from page 27 that about 5 per cent of the composition of the earth's crust is iron, so the prospect of iron starvation is not serious. Nearly every year sees some valuable ore deposit being found. Brazil for example has iron ores estimated at 50,000 million tons now available, with additional reserves in the Amazon region and in the Mato Grosso.

Four regions in the world each make over 100 million tons of

iron and steel per annum and require enough iron ore to manufacture that amount of metal. The western European group of
countries include several with major iron and steel industries:
West Germany, Great Britain, France, Italy, Belgium. France is
one of the biggest ore producers in the world; the other countries,
though containing iron ores, rely to a greater or lesser extent on
imports.

U.S.A. provide two thirds of their own iron ore and obtain one
third from Canada, which is a large exporter. The U.S.S.R. and
eastern Europe get their supplies, appropriately, from inside
the Iron Curtain. U.S.S.R. is the biggest iron ore producing
country and has the world's largest mine at Lebedin, near
Kursk. This contains 20,000 million tons of rich iron ore with
between 45–65 per cent metal content, plus about 10 million
million tons of poorer ore. The U.S.S.R. also has a massive iron
ore deposit near Murmansk by the Finnish frontier, which may
become a useful source of imported iron ore for Britain within
the next few years. Japan is a large importer of iron ores, able to
bargain keenly, with long-term contracts, because of the vast
orders which she can place, with Australia, New Zealand, Chile
and other countries.

The economics of ore exploitation are linked with many
factors, including political ones. Naturally iron producers go for
the richest ores first. As new ore deposits start bringing profit, the
time comes when the country containing the ore wishes to gain
more profit and give more employment to local labour, so they
may instal plants for concentrating, sintering, pelletizing or
otherwise enriching their ores. Economic transport requires
large vessels; Japan for example is developing ore carriers of
over 120,000 tons.

Britain requires over 35 million tons of iron ore each year; less
than half of this is quarried here, mainly in Northamptonshire
and Lincolnshire. The average iron content of all home ores is
now about 27 per cent – rather low by the standards of many
ore deposits elsewhere. Our iron ore imports come from Canada,
Venezuela, Australia, Liberia, Mauretania, Brazil and Sweden,
which ships the iron ore from Norwegian ports. In order to keep
the cost of transport low, deep ports are needed for loading and

unloading. Narvik and Kirkenes in Norway, Port Cartier and Seven Island in Canada are well known, but Puerto Ordaz in Venezuela, Port Monrovia in Liberia, Vitoria da Tubarão in Brazil and Port Etienne in Mauritania owe their development to the millions of tons of iron ore that the world requires.

New Zealand was responsible for an interesting technical development in the transport of ores. For hundreds of miles along the coast of North Island black sand contains iron oxide, much needed by Japan for her flourishing iron industry. The Marcona process involves piping the sand as a slurry into a ship's hold, de-watering it for the voyage, then turning it back into slurry for pumping out of the ship at its destination. The operation is so mechanized that handling a million tons of ore per annum requires only fifty men. Geologists estimate that there are 300 million tons of iron ore sand at Tahatoe, south of Auckland, and probably a similar amount at Waipipi, near Waverley.

PRODUCTION OF IRON IN THE
BLAST FURNACE

Figure 5 is a diagram and Plate 4 an illustration of a British blast furnace. A medium sized furnace is about 30 metres high and 10 metres diameter at its widest part. It is covered on the outside with steel plates 40 mm thick, some of them water cooled, enclosing a refractory lining nearly a metre thick. The hearths of blast furnaces used to be of firebrick but nowadays carbon blocks are used. The correct mixture of ore, coke, and limestone fills the interior of the furnace. A blast of hot air is blown in through a number of nozzles called tuyères (pronounced tweers). There are between twelve and twenty-four of these distributed evenly round the circumference but placed about $2\frac{1}{2}$ metres above the bottom of the furnace in order to leave a space in which the molten iron and slag can accumulate.

The ore, coke, and limestone are given the collective name of the 'charge'. The method of feeding the charge into the furnace is somewhat complicated, the object being to avoid wasting any of the blast-furnace gas and to distribute the contents of the

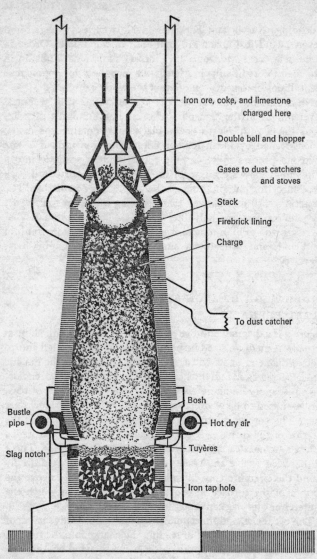

Iron ore, coke, and limestone
charged here

Double bell and hopper

Gases to dust catchers
and stoves

Stack

Firebrick lining

Charge

To dust catcher

Bosh

Hot dry air

Tuyères

Iron tap hole

Bustle
pipe

Slag notch

Fig. 5. Blast furnace

furnace evenly. The charge is hoisted to the top of the furnace and dumped on to a double bell-and-hopper arrangement, as shown in the left-hand drawing in *Fig. 6*. The small bell is then lowered, as in the right-hand drawing, dropping the charge into a lower chamber, at the bottom of which is a larger bell and hopper. When the top has been closed, thus sealing the furnace, the larger bell is lowered and allows the charge to fall into the furnace. The bell is rotated through an angle every time it is used, in order to distribute the charge evenly.

As the process continues, iron is being made from the ore and then melted by the heat of the furnace. Words like 'trickle' or even 'fall' are rather inadequate to describe the conditions in which the enormous contents of the furnace move downwards at the rate of about 500 tons per hour, while the blast of air rushes upwards, the ore being converted into iron and slag, which become melted and which accumulate at the bottom of the furnace. During its descent the iron absorbs a number of impurities from the coke, ore, and limestone, so that the pool of

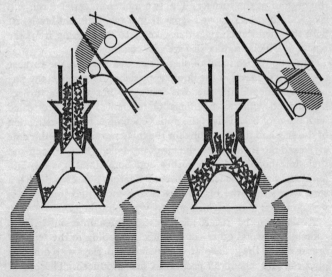

Fig. 6. Bell and hopper for charging the blast furnace

iron at the bottom of the furnace contains about 3 to 5 per cent carbon, about 1 per cent manganese, up to 3 per cent silicon, and usually a small amount of sulphur. In some types of iron, phosphorus is present up to 2 per cent; in others the phosphorus content is only a fraction of 1 per cent.

Most iron ores contain 60 to 80 per cent iron oxide, mixed with silica sand, which is silicon oxide, and earth, clay and rocks which contain silica combined with other compounds. If such a mixture were smelted in a blast furnace, a great deal of the iron would combine chemically with the silica; this was what happened in primitive iron-smelting furnaces. It is therefore necessary to introduce another material which will combine with silica, thus preventing it from uniting wastefully with the iron. Limestone has this property and at the temperature of the blast furnace it forms chemical compounds with the earthy material, including the silica, the result being known as slag. This molten substance descends down the blast furnace; being lighter than iron, it then floats above the layer of molten metal which is accumulating at the bottom of the furnace. The two molten materials, iron and slag, can then be removed separately at two different levels. As with the chemical changes that take place in smelting iron, the reactions which produce slag are very complex.

The pool of iron and of slag grows and rises up the furnace and, when the level of the liquid approaches the tuyères, the slag is drawn off and the metal is 'tapped'. For this the furnace is provided with two openings, or tap holes, which are sealed with plugs of baked clay; the upper opening is for the extraction of the slag and the lower one for the molten iron. When the time arrives to tap the furnace, the lower tap-hole is broken open and the white-hot liquid iron gushes out. Sometimes it is run into moulds, forming the familiar bars of pig iron, but often there is a steel works near the blast furnaces and the liquid iron is conveyed molten to be purified and converted into steel.

A slag can be regarded as a calcium–aluminium silicate, containing several other compounds. According to the types of ores used and the grade of iron manufactured, slags with different physical and chemical properties are produced. These find various uses such as railway ballast, tar-macadam, concrete,

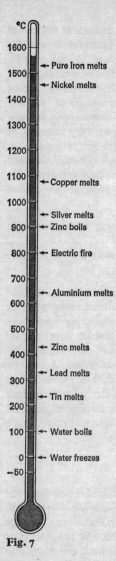

°C

1600   ← Pure iron melts

1500   ← Nickel melts

1400

1300

1200

1100   ← Copper melts

1000

 900   ← Silver melts
  ← Zinc boils

 800   ← Electric fire

 700   ← Aluminium melts

 600

 500

 400   ← Zinc melts

 300   ← Lead melts

 200   ← Tin melts

 100   ← Water boils

   0   ← Water freezes

−50

Fig. 7

porous slag bricks for building, fertilizers, and slag wool for heat insulation. These do not absorb all the available slag and some of it has still to be dumped.

THE HOT BLAST

The primary function of the air blast is to enable the coke to burn and produce a very high temperature. Secondly, it combines with the burning coke to form gaseous carbon monoxide which itself reacts with the iron ore, producing iron and carbon dioxide. The interior of the furnace is hottest at a point just above the level of the tuyères, where the temperature is about 1,800°C.* Between that zone and the top of the furnace, the temperature falls off gradually to about 250°C.

Until about 150 years ago, cold air was blown into the furnace; indeed until recently there were several 'cold-blast furnaces' making a special type of high-grade pig iron. When a cold blast of air was used, as much as eight tons of coal or coke were needed to produce one ton of pig iron. In 1828, a Scot, James Neilson, developed the use of a heated blast and, as a result, the fuel consumption was reduced to five tons per ton of iron. Those who wish to pay tribute to this great metallurgist of the past can find an obelisk to his memory on a hilltop near Kirkcudbright.

---

* That is nearly 300°C higher than the melting point of iron. See *Fig. 7*.

Fifty years ago the hot gas was allowed to burn at the top of the furnace, but now it is not wasted in this manner; it provides an excellent method of pre-heating the blast. On leaving the furnace this gas contains 20–25 per cent carbon monoxide, which is combustible. About 14 to 30 per cent of the furnace gas is used for heating the air blast. This is done by means of 'stoves', usually three to each furnace. These stoves are nearly as large as the blast furnace itself; they consist of, firstly, a chamber in which the carbon monoxide of the furnace gas is burnt and, secondly, a labyrinth of firebricks, arranged in such a way that the burning gases in passing through them heat the firebricks, making them red hot. In operation, two processes are carried out at the same time in the stoves:

(1) two stoves are being heated in the way which has been described above;
(2) the other stove, having been heated, raises the temperature of the blast of air as it passes to the furnace.

By this method, the air is at a temperature between 800–1,200°C when it reaches the blast furnace. About every half hour these operations are changed; the stove which previously was 'on blast' is heated, while one of the stoves which were 'on gas' now gives up its heat to the incoming blast of air. Normally the cycle is half an hour on blast and one hour on gas. As will be seen from *Fig. 8* the gas is passed through dust extractors before being led to the stoves. The remainder of the valuable furnace gas is used for other purposes – for driving pumping-engines and for making sufficient electricity to provide light and power in the works and often for a neighbouring town as well.

MODERN BLAST FURNACES

From the small furnaces of several hundred years ago, the modern blast furnace, with all its ancillary plant, has developed. One of the most impressive achievements in metallurgy has been the growth in size and increase in efficiency of the iron blast furnace, comparable with the difference between the *Mayflower* of 180 tons and the *Queen Elizabeth 2* of 65,000 tons.

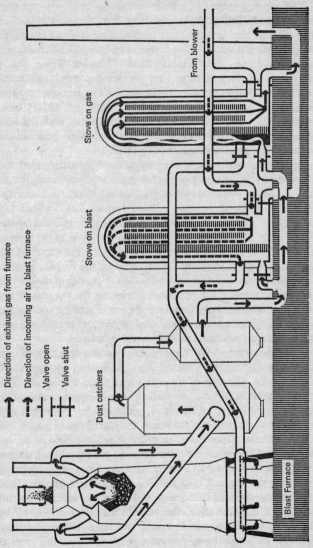

Direction of exhaust gas from furnace

Direction of incoming air to blast furnace

Valve open

Valve shut

Stove on gas

Stove on blast

Dust catchers

From blower

Blast Furnace

Fig. 8

The average output of a blast furnace in 1788 was 912 tons per annum. In Britain the average blast furnace makes more than that tonnage per day – and our performance is small compared with the 'first division' blast furnaces in U.S.S.R., U.S.A., and especially Japan. The Nippon Steel's Nagoya Works No. 3 blast furnace made a record on 27 October 1970, producing 10,080 tons of iron in twenty-four hours. The furnace broke its own record on 1 November when it produced 10,095 tons. Nippon Steel achieved this enormous production with the help of about 200 cubic metres per minute of oxygen injected into the 6,000 cubic metres per minute of air blast.

Japan's development of very large blast furnaces illustrated a principle that was so logical that other iron-producing countries probably wished they had realized it before. An important economic factor in running a blast furnace, and indeed any process, is to achieve maximum output with minimum breakdown time. The really costly breakdown in a blast furnace involves relining, which is made necessary by the attack of the molten cascade of iron and slag descending the furnace and the red-hot air blast hurricaning up it. This is related to the surface area of the furnace lining; but quadrupling the volume of any container involves only doubling the inner surface area so the bigger blast furnace will be expected to give less breakdown cost per ton of iron produced than the smaller furnace.

A medium-large blast furnace that makes 2,000 tons of iron every twenty-four hours uses about 3,500 tons of iron ore, 1,400 tons of coke, 300 tons of limestone and 5,000 tons of air. The furnace itself is the central feature of a complex organization for the production of iron and, although by its height and massiveness it dominates the scene, the blast furnace is only one link in a chain of processes. In order to achieve maximum production and minimum coke consumption per ton of iron, the ore must first be prepared. This usually consists of crushing and screening the ore and then roasting it with limestone and with coke in the form of fine particles known as 'breeze'. In this process of agglomeration, the volatile matter is driven off by the heat, the limestone flux and the coke are incorporated with the ore, and the agglomerate is produced in the form of irregular clinkery lumps, known as 'sinter'. Agglomeration produces a strong material which can be smelted efficiently in large blast furnaces; untreated ore would be crushed into dust by the great weight of the charge and would hamper the passage of gases through the furnace.

In 1960 about 40 per cent of the 'burdens' of British blast furnaces consisted of sinter; by 1974 the proportion had risen to 58 per cent. There is still some way to go before the agglomerating capacity of blast furnace plants in Britain is adequate, but it is sometimes not worth while building sinter or pelletizing plants for those blast furnaces which are due for early replacement. A trend is developing for iron ore to be agglomerated at the orefield itself, which prevents losses due to powdering during transportation.

Plant must be provided to carbonize the coal to coke; to pump and heat the air blast, to receive, store, and charge into the furnace the coke, ore, sinter and limestone flux; to clean and handle the vast quantities of the gas which exits from the blast furnace, and which can be used as a gaseous fuel, and to handle and prepare for sale or disposal the slag which is produced.

Britain has an annual production of about 14 million tonnes* of pig iron per annum, made by about forty blast furnaces. Thanks to technical improvements our output per furnace has gone up, as the following table shows, but for the present we are still in the second division and 1974 showed a decline.

In the search for greater efficiency, Britain is also developing

Table 3

Year	Tonnes per annum per furnace
1938	69,700
1957	147,800
1960	188,400
1965	267,100
1970	316,000
1973	374,400
1974	347,600

* See page 15 for a comment about tonnes versus tons.

Table 4. Pig Iron Production
(*thousands of tonnes per annum*)

	1966	1968	1970	1972	1975
U.S.A.	83·60	81·04	83·30	81·10	73·78
U.S.S.R.	70·26	78·79	85·93	92·30	102·35
Japan	32·02	46·40	68·05	74·06	86·62
Britain	15·96	16·70	17·67	15·32	12·02
France	15·59	16·45	19·22	19·00	17·93
F.R. Germany	25·41	30·31	33·63	32·00	30·03

what might be described as Jumbo blast furnaces. The No. 3
furnace at the British Steel Corporation's Llanwern Works,
commissioned in 1976, has a capacity for 5,000 tonnes of iron
per day. Soon the Corporation will commission a 10,000 tonnes
per day furnace, forming part of the new steelmaking complex
being developed at Redcar, on the south bank of the Tees. With
increased efficiency and the building of larger furnaces, the fuel
consumption has been gradually reduced until today it is very
much less than a tonne of coke per tonne of iron. *Tables 4* and *5*
show firstly the total production of pig iron from six countries,
and then their coke consumption expressed as kilograms of coke
per tonne of pig iron.

The high efficiency of the Japanese furnaces is obtained partly
by injecting oil into the furnace and partly by adding oxygen to
the air blast. Nippon's Tobata Works recorded a monthly

Table 5. Coke Consumption
(*expressed as kilograms of coke per tonne of pig iron*)

	1967	1969	1971	1973
Japan	496	492	448	436
U.S.A.	631	623	624	595
Britain	657	650	610	577
F.R. Germany	604	564	521	495
France	696	649	594	558
U.S.S.R.	601	581	566	558

average coke ratio of 357 kg per tonne in March 1970. Another development, pioneered in the U.S.S.R., has led to the creation of higher gas pressures in the tops of blast furnaces, giving greater efficiency in the utilization of the fuel.

NEW DEVELOPMENTS IN IRON MAKING

Every industry, large or small, needs to take a look at itself fairly frequently. Labour costs increase, fuel or raw materials become scarce due to supplies drying up or become plentiful due to new finds; capital costs mount; new processes are discovered. The trouble very often is that, soon after an industry has spent a great deal of money in capital equipment, some new factors arise. This is particularly true in the iron and steel industry. If you have just spent £300 million on a new iron and steel works the prospects of scrapping it are painful. Blast furnaces have undoubtedly increased in efficiency but not even the Japanese are particularly pleased with their profitability. In the meantime coke, on which blast furnaces have depended for over 200 years, is becoming more expensive, and supplies of suitable coking coals are becoming scarcer.

Many iron-makers consider that the time has come to reduce iron from the ore in a form that can be converted directly into steel. This was experimentally begun in the early 1950s but was not pursued energetically because the improvement of blast furnace efficiency was coming into its stride, so direct reduction process developments were put off till later. As often happens, the developments came from an unexpected quarter. In Mexico, Hojalata y Lamina has been quietly developing a revolutionary process for direct reduction. Already five HyL plants are making enough iron to support a million ton per annum steel-making industry in Mexico.

The ideal for which iron and steel makers are seeking is the reduction of iron ore to a spongy form of iron which can next be converted into steel, probably in electric furnaces. Most of the direct reduction processes have the function of removing oxygen from the iron ore, but not melting the iron. At present there are at least nine processes. Some reduce the iron ore in a rotary kiln using solid fuel or gas; some, like the HyL process, reduce the iron ore to metallic pellets by natural gas in a static hearth. In other processes the iron ore is reduced by carbon monoxide or hydrogen in a shaft furnace. This is a development of the methods which have been used for the last thirty years for making iron powder as discussed in chapter 20.

These direct reduction processes are economic in comparatively small units. They often link up with the production of small or medium sized steel works and they can be operated in countries which are not yet highly industrialized and which therefore do not possess the vast amount of scrap iron and steel on which the major steel-making countries depend.

4
MAKING
ALUMINIUM

Aluminium is the most plentiful metal in the earth's crust, of which it forms eight per cent. The pure metal does not occur in nature; the principal ore from which aluminium can be extracted economically is bauxite, which consists of aluminium oxide combined with water and containing impurities in the form of silica (sand) and oxides of iron and titanium; the iron oxide gives bauxite a red colour. Bauxite is named from the district of Les Baux, near Arles in the south of France, where this important ore was first worked commercially. There are also valuable deposits in the U.S.A., the U.S.S.R., Jamaica, Surinam, Guyana, Australia, Greece, Yugoslavia and Hungary.

Jamaica produced no bauxite before 1952 but production was developed to 7·7 million tons in 1963 and 15 million tons in 1974. Up to December 1969 about 81 million tons of bauxite had been extracted, and another 23 million tons converted and exported in the form of 9·6 million tons of alumina. It is thought that Jamaica has 300 million tons of bauxite. Even greater benefits may come as they develop the smelting of aluminium in the Caribbean region, based on nuclear power.

During the past few years Australia has become an Aladdin's cave of almost every mineral that matters; the continent has been called the biggest quarry in the world. The 250-mile long Darling Range in Western Australia, Weipa in the northern tip of Queensland and Gove in the Northern Territory contain bauxite deposits totalling about 3,500 million tons – ten times those in Jamaica.

Bauxite deposits generally lie near the surface; most are located near the Equator in areas where hot sun and heavy rain have weathered the ore over millions of years. After the vegetation and top soil have been removed, the over-burden of sand or clay is stripped away, either by mechanical scrapers or by powerful hydraulic jets, and the bauxite is mined by open-cast methods. Instead of removing the impurities by slagging during smelting, as in the production of iron, the aluminium ore has to be purified before it can be converted into metal. The concentration process, devised by Karl Josef Bayer in 1890, depends on the fact that aluminium hydroxide (which may be regarded as aluminium oxide combined with water) dissolves in heated caustic soda but the impurities do not; this enables practically pure aluminium oxide (alumina) to be separated eventually.

The ore is first mixed with caustic soda solution in heated pressure chambers. The aluminium hydroxide dissolves in the caustic soda, forming sodium aluminate, which, being soluble, can be passed through filters leaving insoluble impurities behind. On cooling, and 'seeding' the solution with fine particles of alumina, crystals of aluminium hydroxide are precipitated. This is heated to a temperature of about 1,300°C which drives off the water and leaves pure aluminium oxide in a form suitable for conversion into the metal.

The electrolytic process uses a powerful electric current to split up alumina into aluminium and oxygen. Unfortunately alumina possesses the very high melting point of over 2,000°C, so that it cannot readily be liquified. However, the conversion of alumina to aluminium and oxygen, by electrolysis, can occur when it is dissolved in some other substance. These were some of the major problems to solve when, towards the end of the last century, Charles Hall in America and Paul Héroult in France were independently experimenting with the electrolytic method. At about the same time, in 1886, the two pioneers* both arrived at the same solution to their problems, by using a mineral known as cryolite, whose chemical composition is sodium–aluminium–fluoride. Hall and Héroult found that a mixture of cryolite with about five per cent alumina would melt at a little under 1,000°C and that an electric current passed through would keep it molten and simultaneously split up the aluminium oxide into aluminium and oxygen.

* Hall and Héroult were both born in 1863. Both died in 1914.

Cryolite is found in Ivigtut, a settlement on the Arksut Fjord in south-western Greenland. It melts easily and was thought by the Eskimos to be a peculiar form of ice. Until the early 1940s aluminium production depended on the availability of Greenland cryolite and in 1940 the Allies established themselves there to safeguard supplies for their war effort. However the production of synthetic cryolite had begun; though at first the manufactured chemical was dearer than natural cryolite, the synthetic material is now generally used.

Another way in which aluminium production differs from iron-smelting is that the metal is not made in bulk in a large furnace, but is produced in a number of comparatively small units. A modern reduction furnace, illustrated in *Fig. 9*, consists of a shallow rectangular steel container lined with refractory bricks, with an inner lining of carbon serving as the negative electrode, the cathode. Direct electric current enters the furnace through the positive electrodes, the anodes, which are bulky carbon blocks suspended from above and dipping into the molten solution of alumina in cryolite, contained within the carbon

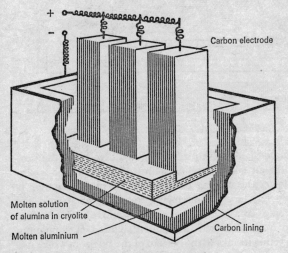

Fig. 9. Extracting aluminium by electrolysis

lining. A modern 100,000-ampère furnace is about 10 metres long and 5 metres wide and produces up to about 750 kg of aluminium every twenty-four hours. Such cells work in series of 150 to 200, called a 'pot line' and rated at about 75 megawatts, using about 15 kilowatt-hours per kilogram of aluminium produced. Carbon paste is put in a mould and is baked by the heat of the furnace as the operation is proceeding. This is known as the Soderburg anode, illustrated in *Fig. 10.*

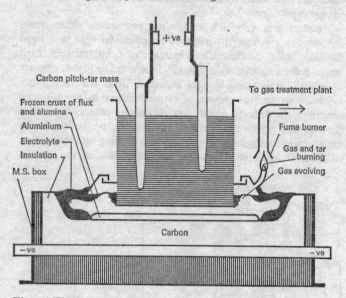

Fig. 10. The Soderburg anode

The alumina is dissolved in the molten cryolite in the proportion of about one in twenty; some calcium fluoride is also added to lower the melting point further. While aluminium is being produced, the cryolite remains practically unchanged, the furnace being continually replenished with alumina. Since aluminium is heavier than the molten solution of aluminium oxide in cryolite, it sinks to the bottom of the cell from which it is periodically sucked out into a travelling ladle, conveyed to a large holding furnace and then cast into ingots. The oxygen liberated at the anodes unites with the carbon to form carbon monoxide which burns at the furnace top, making carbon dioxide. For this reason the anodes are progressively fed into position, as shown in *Fig. 10*. This accounts for the heavy wastage of electrodes, of which about half a tonne is burnt away for each tonne of aluminium produced. The passage of the strong electric current through the cell results in the evolution of heat in much the same way as passing electricity through the elements of an electric fire develops heat. This is sufficient to maintain the minerals in a molten state and to bake the anodes.

The ingredients needed to make a tonne of aluminium illustrate that the smelting of such a metal is a gigantic exercise in mass-handling. Thus $4\frac{1}{2}$ tonnes of bauxite are needed to produce 2 tonnes of alumina by the Bayer process, requiring one tonne of fuel oil and 160 kg of caustic soda. In the electrolysis 15,000 kWh electricity, 450 kg of pre-baked anodes and 50 kg of cryolite are required.

The economic production of aluminium depends largely on the availability and cost of electric power. Up till recently aluminium reduction plants have been associated with hydro-electric undertakings, often in remote mountainous regions. Canada, U.S.A., U.S.S.R., Norway and Switzerland were, and still are, large producers because of their vast hydro-electric resources. Britain unfortunately is short of such power and till recently our production, from the Scottish Highlands, was limited to about 30,000 tons per annum.

The vast capital cost of building dams and hydro-electric power stations, usually in inaccessible places, the hazards of capital investment in some countries with suitable water supplies, and the high costs of transportation from out-of-the-way smelters to manufacturing facilities close to the markets for aluminium products, has altered the economic balance during recent years. Three new projects have been initiated in the United Kingdom and already supplies are becoming available. British Aluminium, owned by the Tube Investment Group, is operating a smelter at Invergordon as a joint operation with the U.S. company, Reynolds. This will have a production of

100,000 tons per annum; the power supply is from the national grid. At Lynemouth in Northumberland, Alcan is developing a plant using coal at a negotiated low price to supply the electric power. The first stage was scheduled to produce 60,000 tons per annum, now increasing to 120,000 tons.

The Rio Tinto Zinc Group, in partnership with British Insulated Callender's Cables and Kaiser Aluminium has a 100,000-ton plant in Anglesey, which obtains electricity from a nuclear power station. The three smelters cost a total of £170 million. Together they will eventually produce 320,000 tons per annum, which when added to the 30,000 tons already obtained by British Aluminium will make Britain nearly self-sufficient for aluminium. In 1974 Britain's aluminium production amounted to 294,000 tons. Even so none of these enterprises will provide electric power as cheaply as in Canada and U.S.A., where a unit of electricity for an aluminium smelter costs much less than in Britain. Here a subsidized power cost of, say, 0.8p per unit would represent £120 per tonne of aluminium and all the other mining and manufacturing costs must be added to that.

The Hall and Héroult method of extracting aluminium uses so much electric power that it is only to be expected that strenuous efforts are being made to discover new and more economic processes.

The Alcoa Company of America is making the first smelting unit, capable of producing 30,000 tons a year of primary aluminium by a new process. Alumina, as already refined from bauxite ore, is combined with chlorine to make aluminium chloride. This is processed in an enclosed cell which separates out molten aluminium and chlorine; the latter is recycled back to the reactor. The new process, with a planned ultimate capacity of 300,000 tons per annum, is expected to reduce the amount of electricity per tonne of aluminium by as much as 30 per cent. Furthermore, it disposes of the need for cryolite.

5
ALLOYS

A motor-car contains many different metals and alloys, chosen because of their properties, the ease with which they can be made into the required shape, their availability and cost. Steel and cast iron make up the greatest weight, with mild steel for the body work, cast iron for many parts of the engine block, and alloy steels in the gears, connecting rods, valves and camshaft. Steel, coated with a layer of aluminium, is used for headlamp reflectors. The automatic gear box is of aluminium alloy, the door handles of zinc alloy, plated with copper, nickel and chromium. Body trim can be chosen from four materials: brightened and anodized high purity aluminium, aluminium-coated plastic, chromium-plated mild steel, or stainless steel. There are small, though vital, components of lesser-known metals and alloys; for example, sparking-plug parts may be of nickel alloy containing manganese, chromium, or silicon; the coil ignition make-and-break contacts are of tungsten. The wheel balance-weights are of lead–antimony alloy.

The majority of the metallic materials in the car are alloys, not pure metals. Practically the only pure metal used in any considerable bulk is copper, in the form of electric wiring, sparking-plug washers, petrol pipes, radiator honeycombs, and the metallic parts of gaskets. A car made solely of pure iron, pure aluminium, and pure copper without the use of alloys, would break down before it had travelled far.

Although pure metals possess some useful properties, such as high conductivity of heat and electricity, they are not often used in structural or mechanical engineering, because their strength is generally insufficient for the arduous duties which materials are expected to perform in our day. The most important way in which the strength of metals can be increased is by alloying. This in turn opens up possibilities of still further improvement of properties by the use of 'heat-treatment', which consists of heating the alloy to a selected temperature, below its melting point, and then cooling it at some pre-determined rate to obtain specially required properties.

An alloy is an intimate blend of one metal, known as the 'parent' or 'base' metal, with other metals or with non-metals. For example, one type of brass contains two-thirds copper and one-third zinc, while the important alloy, steel, is iron containing carbon. Five, ten and fifty pence pieces are made of a copper-nickel alloy.

Several thousand alloy compositions are in commercial use; most are made to standard specifications and widely used throughout industry. Organizations such as the British Standards Institution do a great deal to rationalize the selection of alloys; they keep under regular review new developments, new alloys and of course the preferences of manufacturers and users.

Nevertheless it would be a dull world if we were forced to use only a limited number of materials and if personal choice were altogether removed. The ideal is to achieve a sensible compromise, remembering that a specified and widely used alloy is likely to be more economical than a patented composition. Though it is not very logical, many people like to give names to alloys. We heard of one alloy known as 'Stair Metal', used by a manufacturer in such large quantities that it was included in their standard specifications. Then it was discovered that the name had been given by the storekeeper during his annual stocktaking; an amount of the metal was located under the stairs in an old department and he classified it as 'Stair Metal'. One would not suggest that the world would be a better place if the whole repertoire of alloy metallurgy were limited to a few hundred materials, although the gain in efficiency of production would be considerable. But we sometimes dream of a metallurgist's Utopia, where alloy compositions could not be patented and where proprietary alloys of unspecified composition were unpopular!

HOW ALLOYS ARE MADE

In medieval times, copper alloys were often produced direct, by the smelting of mixed ores of copper and tin, or of copper and zinc. Even nowadays, some alloys are made direct by smelting; thus that complex alloy, cast iron, is the product of the iron blast furnace; 'Monel', a corrosion-resisting alloy of nickel and copper, is made by smelting a nickel–copper–iron ore mined near Sudbury in Canada. However, the composition of most alloys has to be controlled to get the best results, and direct production by smelting mixed ores is, as a general rule, undesirable.

Most alloys are prepared by mixing metals in the molten state; then the mixture is poured into metal moulds and allowed to solidify. Generally the major ingredient is melted first; then the others are added to it and should dissolve completely. For instance, if a plumber makes solder he may melt his lead, add tin, stir, and cast the alloy into stick form. Some pairs of metals do not dissolve in this way and when this is so it is unlikely that a useful alloy will be formed. Thus if the plumber were to add aluminium, instead of tin, to the lead, the two metals would not dissolve – they would behave like oil and water; when cast, the metals would separate into two layers, the heavy lead below and aluminium above.

One difficulty in making alloys is that metals have different melting points. Thus copper melts at $1,083°C$, while zinc melts at $419°C$ and *boils* at $907°C$; so, in making brass, if we just put pieces of copper and zinc in a crucible and heated them above $1,083°C$, both metals would certainly melt, but at that high temperature the liquid zinc would boil away and the vapour would oxidize in the air. The method adopted in this case is to heat first the metal having the higher melting point, namely the copper. When this is molten, the solid zinc is added and is quickly dissolved in the liquid copper before very much zinc has boiled away. Even so, in the making of brass, allowance has to be made for unavoidable zinc loss which amounts to about one part in twenty of the zinc.

Sometimes the making of alloys is complicated because the

metal with the higher melting point is in the smaller proportion; for example, one light alloy contains $95\frac{1}{2}$ per cent aluminium (melting point 659°C) with $4\frac{1}{2}$ per cent copper (melting point 1,083°C); to manufacture this alloy it would be undesirable to melt the few kilograms of copper and add over twenty times the weight of aluminium. The metal would have to be heated so much to persuade the large bulk of aluminium to dissolve that gases would be absorbed, leading to unsoundness. In this, as in many other cases, the alloying is done in two stages. First an intermediate 'hardener alloy' is made, containing 50 per cent copper and 50 per cent aluminium, which alloy has a melting point considerably lower than that of copper and, in fact, below that of aluminium. Then the aluminium is melted and the correct amount of the hardener alloy added; thus, to make 100 kg of the aluminium–copper alloy we should require 91 kg of aluminium to be melted first and 9 kg of hardener alloy to be added to it.

Not all hardener alloys are of fifty-fifty composition; for example, when nickel is alloyed with aluminium it is added in the form of a nickel–aluminium alloy containing 25 per cent nickel. In the making of tool steels, cutlery steels, and other alloy steels, manganese, chromium, or silicon are added as ferro-manganese, ferro-chrome, or ferro-silicon, which incidentally are cheaper to produce than the pure elements. ('Ferro' indicates the presence of iron; thus ferro-chrome contains 20 to 55 per cent iron and 80 to 45 per cent chromium.) Making up an alloy to an exact composition is not easy and the melter is therefore allowed a specified working margin as regards both major elements and impurities. The composition of the alloy is checked by analysis, after which it is released for use in the foundry, rolling mill, or forge.

THE MELTING POINT OF ALLOYS

In winter time, when ice forms on the roads, salt is thrown down to 'melt' the ice. This practical use of a natural phenomenon demonstrates that a mixture of salt and water has a lower freezing point than that of pure water. The temperature at which pure water just freezes is not low enough to cause the mixture of water and salt to freeze; therefore, providing the temperature is not too low, the salted ice melts. The same effect is to be found in metallurgy, for when one metal is alloyed with another, the melting point is always affected.

In a few cases, the melting point of the alloy can be worked out approximately by arithmetic. For instance, if copper (melting point 1,083°C) is alloyed with nickel (melting point 1,454°C) a fifty-fifty alloy will melt at about halfway between the two temperatures. Even in this case the behaviour of the alloy on melting is not simple. A copper–nickel alloy does not melt or freeze★ at one fixed and definite temperature, but progressively solidifies over a range of temperature. Thus, if a fifty-fifty copper–nickel alloy is melted and then gradually cooled, it starts freezing at 1,312°C; as the temperature falls, more and more of the alloy becomes solid until finally at 1,248°C it has solidified completely (*Fig. 11*). This 'freezing range' occurs in most alloys, but it is not found in pure metals, metallic or chemical compounds, nor in some special alloy compositions, referred to below, which melt and freeze at fixed temperatures.

The alloying of tin and lead furnishes an example of one of these special cases. Lead melts at 327°C and tin at 232°C. If lead is added to molten tin and the alloy is then cooled, the freezing point of the alloy is found to be lower than the freezing

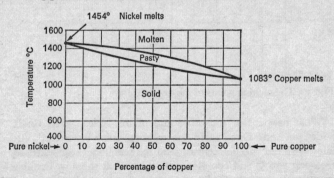

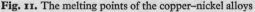

Fig. 11. The melting points of the copper–nickel alloys

★A liquid metal is said to 'freeze' when it becomes solid. Thus copper freezes at 1,083°C, just as water freezes at 0°C.

points of both lead and tin (see *Fig. 12*). For instance, when a molten alloy containing 90 per cent tin and 10 per cent lead is cooled, the mixture reaches a temperature of 217°C before it begins to solidify. Then, as the alloy cools further, it gradually changes from a completely fluid condition, through a stage when it is like porridge, until it becomes as thick as paste, and finally, at a temperature as low as 183°C, the whole alloy has become completely solid. By referring to *Fig. 12* it can be seen that with 80 per cent tin, the alloy starts solidifying at 203°C and finishes only when the temperature has fallen to 183°C (note the recurrence of the 183°C).

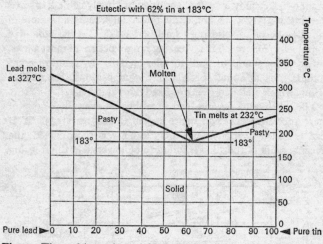

Fig. 12. The melting points of the tin–lead alloys

What happens at the other end of the series, when tin is added to lead? Once again the freezing point is lowered. An alloy with only 20 per cent tin and the remainder lead starts to freeze at 276°C and complete solidification at the now familiar temperature of 183°C. The tin–lead alloys within the range 19·5 to 97·4 per cent tin finish freezing at the same temperature of 183°C. One particular alloy, containing 62 per cent tin and

38 per cent lead, melts and solidifies entirely at 183°C. Obviously this temperature of 183°C and the 62/38 per cent composition are important in the tin–lead alloy system. Similar effects occur in many other alloy systems; the special composition of each series which has the lowest freezing point and which entirely freezes at that temperature is known as the 'eutectic' alloy; the freezing temperature (183°C in the case of the tin–lead alloys) is called the eutectic temperature. The word is derived from the Greek *eutektikos* meaning 'capable of being easily melted'.

By a careful choice of constituents, it is possible to make alloys with unusually low melting points. Such a fusible alloy is a complex eutectic of four or five metals, mixed so that the melting point is depressed until the lowest melting point possible from any mixture of the selected metals is obtained. A familiar fusible alloy, known as Wood's metal, has a composition by weight:

bismuth	4 parts
lead	2 parts
tin	1 part
cadmium	1 part

and its melting point is about 70°C; that is, less than the boiling point of water. Practical jokers can amuse themselves by casting this fusible alloy into the shape of a teaspoon, which will melt when used to stir a cup of hot tea. Incidentally, the bismuth is metallic bismuth and not the medicinal bismuth carbonate, usually called 'bismuth' and taken as a palliative for indigestion.

These low-melting-point alloys have been used for antifire sprinklers. Each jet contained a piece of fusible alloy which melted and released water if a fire occurred and the temperature of the room rose.

THE STRENGTH OF ALLOYS

If sugar is mixed with sand, the mixture displays the properties of both ingredients; it is sweet but gritty and the more sand in the sugar the less pleasant is the taste. Metals rarely behave in such a predictable manner. When one metal is added to another, the alloy often has a new individuality which one would not expect from the properties of the two metals.

The laws of heredity teach us that unhealthy parents are likely to beget weakly children, but a weak or soft metal alloyed with another weak element may produce a strong alloy with strikingly different properties from those of the parent metals. For example, pure iron is a soft metal and carbon in its commonest form is mechanically weak, but as little as half of 1 per cent carbon in iron produces a strong steel which responds to hardening and tempering treatments, and which can be used for making railway lines or rifles.

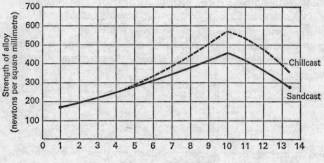

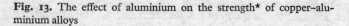

Percentage of aluminium

Fig. 13. The effect of aluminium on the strength* of copper–aluminium alloys

Again, copper and aluminium are both fairly weak. But the addition of 5 per cent aluminium produces an alloy twice as strong as copper; with a 10 per cent aluminium content, the 'aluminium bronze' produced is three times as strong. *Fig. 13* shows a graph representing the increase of strength of copper alloys obtained by adding aluminium. It will be seen that somewhat higher strength is obtained when the alloy is rapidly cooled, or chill-cast, in metal moulds.

Naturally, the temptation would be to go on adding more and more aluminium in the hope of getting still more spectacular

* See pages 15 and 116 for a discussion of the new SI units.

increases of strength. Unfortunately, this does not occur. With 10 per cent aluminium, the aluminium bronze is as strong as mild steel, very tough, and in many ways a valuable engineering alloy. At 13 per cent aluminium, the properties have already deteriorated, and at 16 per cent aluminium, the alloy is about as brittle as a carrot and much less useful. There is a no-man's land between 16 and 88 per cent of aluminium where the alloys are valueless for engineering purposes. (The fifty-fifty hardener alloy mentioned on page 62 is certainly used for making up alloys where its relatively low melting point and its brittleness are convenient. It is, however, only an intermediate product, not an engineering material.) At the other end of this copper–aluminium series, the alloys rich in aluminium, containing up to 12 per cent copper, are again of service. The high strength of aluminium bronze is not attained but the aluminium-base alloys containing copper are considerably stronger than pure aluminium and they, too, are used in industry. These alloys are nearly as light as pure aluminium and this is a special reason for their importance.

It is rare for *all* ratios of two metals to have useful properties, though there are exceptions. For instance, the copper–nickel alloys are valuable throughout practically the whole range of composition.

By blending suitable metals it is possible to produce alloys whose properties are quite different from those of the ingredient metals. The melting point of the alloy may be different from, and in some cases lower than, that of either of the parent metals. Alloys may be considerably harder and stronger than pure metals. Furthermore, as will be shown later, many of these alloys respond to heat-treatment processes, whereby a still wider range of useful properties is attainable. In this class come the steels as well as the strong 'light alloys'. Thus, whenever some special property is required – great strength, toughness, resistance to wear, high electrical resistance, magnetic properties, or corrosion resistance, it is usually possible to produce it in a carefully chosen, skilfully made, and intelligently manipulated alloy.

6
METALS UNDER
THE MICROSCOPE

The microscope is the most useful scientific instrument which
assists the metallurgist in his everyday work; with its aid a
skilled observer can learn a great deal about the structure, the
manufacturing history, the effect of heat-treatment, and the
causes of breakage of any piece of metal or alloy. The microscope
as used in a works laboratory is generally capable of magnifying
up to 1,000 times, and by the use of different objective lenses a
metal can be examined under several successive magnifications.
Usually a camera is provided so that the structure of a metal can
be photographed as well as observed visually.

As metals are practically opaque, the metallurgical microscope
is designed to examine their surfaces by reflected light. A specimen of convenient size, about 10 millimetres cube, is cut off.
One face is filed flat and then ground with successively finer
grades of carborundum papers which are lubricated with paraffin
or water and are supported on plain glass surfaces. Final polishing for optical viewing is continued on cloth into which is
worked a slurry of fine alumina or magnesia, to produce a mirror-
like finish on the metal.

First the polished surface is examined under the microscope
to see if there are any cracks, inclusions, or holes. *Fig. 14a* shows
the appearance of a piece of polished wrought iron; the dark
streaks are inclusions of slag which are always present in this
metal. Copper of certain grades may contain small particles of
copper oxide which can be identified microscopically. If a metal
has failed in service, the position and shape of the crack can be
examined and the cause of the failure may be diagnosed; for
example, it may be found that the crack began at a large piece of
slag or other inclusion.

Such an examination, useful though it is, does not indicate the structure or composition of an alloy, and it tells little of the success or otherwise of heat treatment. Much information can, however, be obtained by etching the polished surface of the metal, and re-examining it under the microscope. The metal is immersed in a chemical solution, chosen according to the metal that is to be examined, and the particular constituent it is desired to observe.

For general examination, steel is usually etched in a mixture of 98 parts alcohol and 2 parts nitric acid, though if one particular constituent, known as 'cementite', has to be identified, the steel is etched in boiling sodium picrate solution which makes the cementite appear black. The time of etching varies from a few seconds to half an hour, depending on the alloy and on the solution employed. Sometimes more sophisticated methods are used, such as electrolytic attack or distilling away some of the metallic surface in vacuum or by slightly heating the specimen to form thin films of oxide.

THE GRAIN STRUCTURE OF METALS

Microscopic examination of the etched surface of a polished metal reveals that it is built up of innumerable small 'grains'. *Fig. 14b* shows the same piece of wrought iron as in *Fig. 14a*, but this time the metal has been etched in a dilute solution of nitric

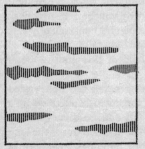

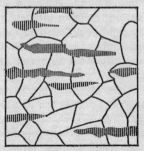

a Before etching   b After etching

Fig. 14. Wrought iron

acid in alcohol. It will be understood that the illustration represents a section cut through the metal and that in reality the grains are three-dimensional.

Like many other technical words 'grain' has a somewhat different meaning from the popular interpretation, which connects grains with sand, salt, or sugar. A grain of metal differs from one of sugar in at least two respects. Firstly, each grain of sugar is brittle and can be crushed into powder, whereas metal grains are usually ductile. Secondly, little cohesion exists between the grains of a lump of sugar, while in a block of metal great force is needed to separate the grains. Sometimes this cohesion of the boundaries may be weakened; for example, when one part in ten thousand of bismuth is present in copper or gold, the bismuth distributes itself at the grain boundaries, thus reducing cohesion and causing the piece of copper or gold to be brittle.

On etching the metal, the grain boundaries are more readily attacked than the interior of the grains. Tiny channels are eaten away at the extremities of each grain; when the etched metal is examined under the microscope, light falling on these channels is scattered, so that the boundaries around the grain appear dark. Under some conditions of etching, whole grains appear contrasted in tone, which may be even more distinguishing than the demarcation of the boundaries. *Fig. 15* illustrates the cause of this contrast. It will be seen that the grains have been attacked by the etching solution in different ways. When light falls on the grain marked '*a*' most of it will be reflected back through the objective of the microscope and consequently this grain appears bright. On the other hand light falling on grain '*b*' is reflected sideways and so the whole of this grain appears darker.

Although the grain structure of most metals can be revealed only after etching and with the aid of the microscope, there are a few examples where large metal grains can be seen with the unaided eye. When brass articles such as door knobs, or the long brass handles at the entrances to public buildings have been in use for a year or so, the constant rubbing with human hands polishes and etches the brass, so that the grains are revealed as small patches or spangles of slightly varying colour tones. In such brass knobs and handles the grain size is comparatively large, of the order of two millimetres. Large grains of zinc can also be seen on galvanized ware.

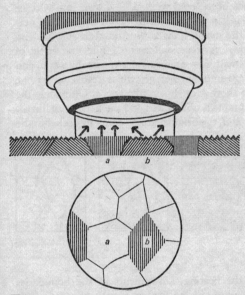

Fig. 15

However, in most metals and alloys the grain structure can be distinguished only by the use of a microscope. For example, some steels, in the condition in which they are used in industry, have grains about one fourth of a millimetre across. The grain size of a metal depends on the casting temperature, the impurities present, and the mechanical working and heat-treatment to which the metal has been subjected; this is one of the important ways in which metallurgists can fit metals for the tasks they will have to perform in service. In general, a metal with fine grains will be somewhat harder and stronger than one with coarse grains. In some instances, the grain size affects the surface appearance of metals. For example, if a piece of brass sheet has unduly large grains, it may show an 'orange peel' effect on the surface, and this makes it so rough that the brass may be unsuitable for the commercial production of articles such as, for example, headlamp reflectors.

After a metal has been 'cold worked' – drawn into wire or rolled into sheet – it is possible to bring about the birth of new grains by heating the metal. This is illustrated in *Fig. 16*, and the process is known as 'recrystallization'. The first effect of heating is to form small, new grains as shown in white in *a*, and these rapidly enlarge until further growth is restricted by one new grain meeting another as shown in *b* and *c*. Ultimately the original system of grains is obliterated and the new, crystallized, structure is shown in *d*, the original grains being indicated in the drawing by dotted lines.

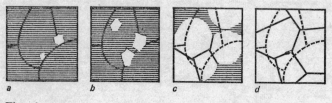

Fig. 16

On continuing the heating, adjustments of the boundaries may take place, resulting in the further growth of some grains at the expense of others. The eventual size of the grains in a piece of metal depends on the amount of deformation previously existing and on the time and temperature of the heating process. By special means, a piece of metal can be treated so that it becomes one large grain, known as a 'single crystal'.

The production and study of single crystals has provided much valuable information to scientists. Often a metallic single crystal possesses remarkable properties of ductility in one or more directions. Some single crystals, diamonds for example, are produced by nature. Single crystals of copper sulphate can be produced from a solution in any laboratory. Large metallic single crystals can be 'grown' by using a small crystal of the metal to 'seed' the vertical withdrawal of a single crystal, which solidifies from the surface of the molten metal. It is also possible to form single crystals in solid, worked, metals by applying 1–5 per cent strain, followed by annealing. Small single crystals can be grown by vaporizing a material in a vacuum and letting it condense slowly.

THE STRUCTURE OF ALLOYS: SOLID SOLUTIONS

The properties of any metal are altered when it is alloyed with another, and this makes one wonder whether microscopic examination will reveal structural differences between pure metals and alloys. In some cases such differences can be observed; thus, two constituents can usually be identified in the lead–tin or the iron–carbon alloys. But other alloys may consist of the polyhedral grains characteristic of a pure metal, and only one constituent is discernible. For example, the microstructure of the copper–nickel alloys is as shown in *Fig. 17*, which may be compared with that of wrought iron in *Fig. 14b*.

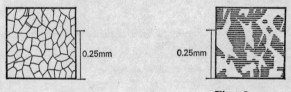

Fig. 17   Fig. 18

If this alloy were composed of grains of copper mixed among grains of nickel, it would be possible to distinguish them by colour alone. Clearly then, something must have happened to the copper and nickel atoms to mingle them so closely that microscopic examination cannot reveal the individual metals.

The two metals are said to exist in a state of 'solid solution'. It may seem strange that one solid metal can exist in solution in another, but there is a wider definition to the word 'solution' than merely 'something dissolved in a liquid'. A solution may be described as an intermingling of one substance in another so closely that the dissolved substance cannot be distinguished or separated by mechanical means. This description can be applied to the condition of the solid copper–nickel alloy just as it is applied to sugar dissolved in water.

Only a few pairs of metals, such as copper and nickel, can exist in solid solution throughout the whole range of possible compositions, but most metals can contain at least some of another metal in solid solution. In an alloy where the metals do not show complete solid solubility, separate constituents may be recognized by examining the etched alloy under the microscope. *Fig. 18* shows the microstructure of a brass containing about 40 per cent of zinc, similar to that used for domestic water taps or brass nuts. The two constituents are two solid solutions, each of different composition.

Just as tea dissolves more sugar when hot than when cold, so a metal can usually retain more of another metal in solid solution when it is hot than when cold. *Table 6* shows the solid solubility of various metals in magnesium at room temperature and at 300°C. It will be seen that the solid solubility of each metal is greater at 300°C.

Table 6. The Solid Solubility of Various Elements in Magnesium at Room Temperature and at 300°C

Element	Solid solubility in magnesium (Weight per cent)	
	Room temperature	300°C
Aluminium	2·3	5·3
Calcium	about 0·1	0·18
Copper	under 0·1	0·1
Lead	3·7	16·0
Manganese	nil	0·1
Silver	1·5	3·6
Zinc	1·7	6·0

INTERMETALLIC COMPOUNDS

In the aluminium–copper alloys, solid aluminium at about 530°C can retain 5 per cent of copper in solution, while at room temperature, it can normally hold less than half of 1 per cent. Therefore, if an alloy containing, say, 4 per cent copper is slowly cooled from about 530°C, it comes to a stage, at just below 500°C, when it can no longer hold as much as 4 per cent copper and, as the temperature falls further, less and less copper can be retained in solution. The surplus copper does not separate as distinct grains of that metal, but in the form of an 'intermetallic compound' to which the symbol $CuAl_2$ is given.*

As the temperature becomes progressively lower, more and more copper comes out of solution and forms $CuAl_2$, so that finally the slowly cooled alloy at room temperature consists of a background of aluminium containing only a small amount of copper in solid solution, together with a number of island grains of the intermetallic compound $CuAl_2$ dispersed throughout the alloy. When isolated, this compound is found to have characteristic properties; it is, for example, extremely hard. Plate 23a shows the microstructure of a lead–tin–antimony alloy, such as is used for bearings. The 'cubes' are of an intermetallic compound of tin and antimony.

The occurrence of intermetallic compounds in alloy systems is frequent – rather the rule than the exception. When the composition of an alloy is such that its structure consists of a matrix of solid solution and particles or grains of an intermetallic compound, the alloy is likely to be useful, for it combines the toughness of the solid solution with the hardness of the intermetallic compound. But if the intermetallic compound predominates, that alloy displays hard and brittle properties and consequently may be unserviceable. This accounts for the fact that only limited ranges of composition of alloys are suitable for engineering; for example, in the alloys formed with aluminium and copper, the useful ones occur within two ranges:

* $CuAl_2$: its composition by weight is 54 per cent copper, 46 per cent aluminium. The formula denotes that one copper atom is intimately associated with two aluminium ones.

(a) The aluminium-rich alloys with 0 to 12 per cent copper;
(b) The copper-rich alloys with 0 to 13 per cent aluminium.

The intermediate alloys are unserviceable in engineering.

EQUILIBRIUM CONDITIONS

The behaviour of the aluminium–copper alloys will be referred to again on page 190 when 'Duralumin' is discussed, but it may be remarked here that the complete separation of constituents is effected only on slow cooling. If the aluminium alloy containing 4 per cent copper is quenched in water from 500°C so that it is rapidly cooled to room temperature, the copper does not have time to come out of solid solution and the aluminium at room temperature is forced to hold more copper in solid solution than it should. In other words it is 'super-saturated' and it may be some days afterwards before the copper atoms spontaneously separate. In some other alloys treated in this way a state can be reached by quenching where adjustment can be attained only by warming the alloy, which gives opportunity for the separation to take place.

Many of the phenomena of metallurgy may be attributed to this sluggishness of alloys in attaining equilibrium. The hardening and subsequent tempering of steel and the age-hardening of duralumin depend on the fact that rapid quenching in water makes the condition of the alloy different from that produced by leisurely cooling in a furnace.

THE STRUCTURE OF EUTECTICS

There is another type of constituent seen under the microscope which may appear in those series of alloys which form eutectics. It will be remembered from page 65 that in certain alloys there exists one particular composition which melts at a lower temperature than any other alloy in that series. This alloy is known as the eutectic. The structure of eutectics may consist of alternate thin layers of the metals concerned, or in other cases small globules of one metal embedded in a matrix of the other.

For example the silver–copper eutectic is composed of 72 per cent silver and 28 per cent copper. If an alloy in such a series is not of eutectic composition, the structure as seen under the microscope consists of one of the metals and some eutectic. Thus an alloy of 10 per cent silver and 90 per cent copper has a structure of grains of copper plus a small amount of eutectic. Similarly at the other end of the series, an alloy with 90 per cent silver and 10 per cent copper consists of eutectic plus silver. The nearer the composition approaches that of the eutectic, the greater the proportion of eutectic seen when the alloy is viewed under the microscope.

THE BIRTH OF GRAINS FROM MOLTEN METALS

Most people have seen the attractive patterns formed in winter when water vapour freezes on windows. A similar beautiful structure may be formed when a liquid alloy solidifies, though the alloy has to be polished, etched, and examined under the microscope to make its structure apparent. When a molten alloy begins to freeze, minute crystals form at various points in the liquid and these start to grow by developing branches in certain directions, as shown in *Fig. 19a* and *b*. These tree-like formations are known as 'dendrites'. When the arm of a dendrite meets that of another, as in *c*, outward growth is restricted, but the spaces between the branches continue to fill in until all the metal is solid, as in *d*. Photographs of dendrites, on antimony, and in a copper–silver alloy, are shown in Plates 22a and 23b.

When a solid piece of *pure metal* is sectioned, polished, etched, and examined under the microscope, no sign of the dendrites

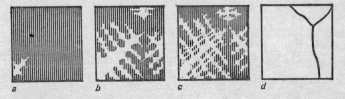

a   b   c   d

Fig. 19

can be seen, because the metal is uniform. But when a cast *alloy* is solidified and examined microscopically, evidence of the dendrites can usually be seen, for the composition of the first part of the alloy to freeze differs from that which finally freezes, and what is known as a 'cored structure' is produced. If such an alloy is then worked either hot or cold and is heat-treated, this 'cored structure' may be gradually eliminated and a homogeneous solid solution may be formed, having the uniform structure which has been shown in *Fig. 17*.

The technique of the microscope and the science of metal structures have made many advances in the last few years; methods of increasing refinement are enabling us to look further and further into the structure of alloys. Thirty years ago a magnification of 1,500 was reckoned as high; then it became possible to photograph structures at magnifications up to 7,000, by the use of ultra-violet light. Now, with the 'electron microscope' (see Plate 24), magnifications up to nearly 100,000 are possible and we are approaching the point at which we can expect to see, if not atoms themselves, at least the effects produced by quite small groups of atoms. Recently a refined instrument known as an 'electron beam scanning micro-analyser' or 'microprobe' has been developed and produced commercially. A beam of electrons is 'scanned' over a microscopic area of metal and generates X-rays which are picked up by a probe and spectrometer. This enables metallurgists to identify the compositions of all constituents and impurities of a sample of metal. By the use of such methods, metallurgists are approaching the ultimate regions in which many basic phenomena of the science of metals may be studied.

7
THE INNER
STRUCTURE OF METALS

All chemical elements, including the metals, are composed of atoms, which are the smallest particles retaining the individual characterstics of the element in question. The atoms are so small* that they cannot be seen, even with the most powerful microscope. Despite this handicap, physicists have been able to find out a great deal about atoms, and this has helped to explain the behaviour of metals. It is important to know how the atoms are arranged in a grain of metal. Are they all piled at random, or do they exist in a regular and orderly formation? Are the atoms of all metals exactly the same size, or does an atom of, say, lead occupy less space than an atom of magnesium?

Even in the latter years of the nineteenth century, scientists had evidence that the atoms in metals were arranged in a regular geometrical fashion, that is, metals belonged to the class of substances which are 'crystalline'. But it was not until 1911 that Max von Laue produced proof of this belief by the examination of metals with X-rays.

The conception that metallic grains are crystals is often a stumbling-block to those who are beginning to study the science of metals. Most people associate crystals with such substances as quartz, diamond, or copper sulphate, which are hard and sparkling, but this does not imply that all crystalline bodies have these characteristics. The basic definition of crystalline nature does not concern the outward appearance of a substance, but arises from its inner symmetry. There are many quite soft substances which are truly crystalline, whilst some hard sparkling substances, such as glass, are not true crystals at all.

* It has been calculated that a cubic millimetre of copper contains about 84,693,000,000,000,000,000 atoms.

Metallurgists and physicists have discovered the different patterns which the atoms adopt in the various metals, the distance between neighbouring atoms, and the space which the individual atoms occupy. In referring to the patterns in which atoms are arranged, metallurgists speak of 'lattice' structures. In aluminium, copper, nickel, lead, silver, gold, platinum, and several other metals, the atoms are spaced evenly in rows at right-angles to each other. This can be likened to atoms arranged at each corner of millions of adjoining cubes, while other atoms occupy positions at the centre of each of the cube faces. This particular atomic lattice pattern is known as 'face-centred cubic', and is illustrated in *Fig. 20*.

In iron, when at room temperature, and in several other metals such as vanadium, tungsten, molybdenum, and sodium, the atoms are disposed in another type of cubic pattern. There is again an atom at each corner of each imaginary cube, but instead of other atoms occurring at the middle of the cube faces, a single atom is located at the centre of every cube. This structure is known as 'body-centred cubic' and is illustrated in *Fig. 21*. Iron is particularly interesting, for at room temperature its atoms are arranged in the body-centred cubic form, but at 906°C the atoms reshuffle into the face-centred cubic pattern, while at a still higher temperature of about 1,400°C the iron atoms change back to a body-centred cubic lattice.

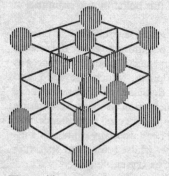

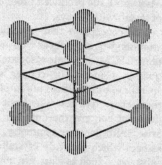

Fig. 20. Face-centred cubic pattern

Fig. 21. Body-centred cubic pattern

In zinc, magnesium and cadmium the atoms are arranged in a hexagonal pattern, while other metals have still more complex lattices. Most of the common metals are either face-centred cubic, body-centred cubic, or hexagonal in their atomic lattices.

After having discovered how the atoms of pure metals were arranged, investigators turned their attention to the lattice patterns of alloys. From microscopic examination they already knew that the grain structure of alloys was often different from that of the metals of which they were composed, and so it seemed likely that some parallel difference might be found when the atomic lattices of alloys were investigated. There were interesting problems which required elucidation; for example, what happens to the atomic arrangement when a metal which crystallizes in the face-centred cubic pattern is alloyed with one of a 'hexagonal' type? Can any light be thrown on the cause of the increase of strength which occurs when one metal is alloyed with another? In the following discussion we consider in some detail the behaviour of one alloy, brass, and then give a summary of various points.

THE ATOMIC LATTICE PATTERNS OF BRASS

The atoms in pure zinc are disposed in a hexagonal arrangement whereas copper atoms form a face-centred cubic lattice structure. The space occupied by the zinc and copper atoms differs, that of zinc being about 13 per cent larger than that of the copper atoms. So when these two metals are alloyed to form brass, some complications may be expected. When only small amounts of the zinc are present in brass, the prevailing atomic arrangement is similar to that of copper, which means that zinc atoms, in solid solution, have to adapt themselves to the face-centred cubic pattern. Each zinc atom takes the place of one copper atom, but because the space occupied by the zinc atom is greater than that of copper, the face-centred cubic lattice is distorted at the point where the 'stranger atom' is introduced, and the whole of the lattice becomes slightly larger. Perhaps a comparison of *Fig. 22* with *Fig. 20* will demonstrate this better than words.

When progressively increasing amounts of zinc are alloyed

with copper, the brass becomes increasingly hard; this can be explained in part by referring to the drawing of the stranger atom. The introduction of a new atom of different size from the rest brings about a condition of distortion in the lattice, and this leads to a greater resistance to deformation than occurs in the pure metal. The strengthening by alloying can occur whether the stranger atoms are larger or smaller than those of the original metal, for distortion of the lattice occurs in either case.

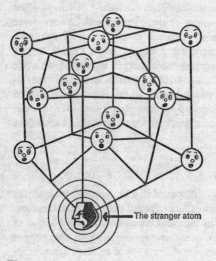

The stranger atom

Fig. 22

As more zinc is added, the face-centred cubic arrangement of the atoms becomes increasingly distorted till, when about 36 per cent of zinc is present, the pattern becomes unstable and here and there another form of atomic lattice comes into existence, which is body-centred cubic. Up to this point the strength and hardness of the brass gradually increase with rising zinc content, but at the 36 per cent composition the properties are sharply altered and further additions of zinc bring about a more rapid rise of hardness than before.

A change can also be noticed when the alloy is examined under the microscope. Up to 36 per cent zinc, the brass is in the form of a solid solution in which no direct indication of the presence of the zinc can be observed except by some change of colour. With over 36 per cent of zinc a new constituent or 'phase' begins to appear and can be seen under the microscope (see *Fig. 18*, page 73). It is necessary to distinguish between these phases, and they are given Greek letters to designate them. According to the usual practice, the first phase is called alpha (α), the next beta (β); a brass containing 36 to 42 per cent of zinc includes both alpha and beta phases, and is called an alpha–beta brass. When still more zinc is added, further phases appear with different atomic lattices; and with each new structure the mechanical properties of the brass alter sharply. In all, there are five different constituents which can exist, though not more than two at once, in the copper–zinc alloys. The effects of increasing amounts of zinc alloyed with copper are summarized in *Table 7*, which compares the composition of the alloys with their microstructure, atomic lattice, and mechanical properties.

The copper–zinc alloys are rather complicated because so many atomic reshufflings occur. Some alloy systems, such as the copper–tin series, are even more complex; others, such as the copper–nickel alloys, are straightforward because both metals have face-centred cubic atomic lattices, their atoms are of similar size, and they are able to form a continuous range of solid solutions, the lattice dimensions changing gradually, from pure copper to pure nickel.

Five points of general interest concerning the atomic structure of metals are given below:

(1) The atoms of each metal have a characteristic size, differing from all the others.

(2) The atoms in each grain of a solid metal are arranged in a regular pattern. There are several types of patterns, the three most important for metals being known as face-centred cubic, body-centred cubic, and hexagonal.

(3) When one metal exists in solid solution in another, the atoms of the added element usually take their place on the

Table 7. Some Changes Produced by Alloying Zinc with Copper

| Composition of alloy | Atomic structure — Description of lattice | Spacing of atoms Angstrom Units | Crystal structure as seen under microscope | Mechanical properties* | | Tensile strength | | Ductility (expressed as per cent on elongation length) |
				Zinc per cent	Diamond pyramid hardness number	Tons per sq. inch	Newtons per sq. mm	
Pure copper	Faced-centred cubic	3·607	Grains of pure copper	—	53	15	230	45
Copper alloyed with up to 36% zinc	Faced-centred cubic pattern which progressively increases in size	3·607 with 0% zinc, increasing to 3·963 with 36% zinc	Grains of solid solution of zinc in copper (known as alpha phase)	10 20 30	60 62 65	18 20 21	280 310 325	55 65 70
Copper with about 36% zinc	A new structure appears (body-centred cubic) in addition to the original face-centred cubic structure	3·693 for alpha constituent. 2·935 for new (beta) constituent	Small quantities of a new constituent appear (known as beta phase)	36	70	22	340	60
Copper with 36 to 42% zinc	With increasing zinc more of the alloy consists of the new constituent having body-centred cubic lattice	On change to body-centred cubic lattice the atoms are 2·935 units apart	The beta constituent increases in quantity while the alpha constituent diminishes	40	85	27	395	45
Copper with 42 to 52% zinc	The lattice is entirely body-centred cubic	2·935 with 42% zinc, increasing to 2·941 with 52% zinc	The structure is entirely beta	45	90	28	410	20

* The units of hardness, tensile strength, and ductility are explained on pages 115 to 121.

atomic lattice of the parent metal, in spite of the fact that the atomic size of the added metal is different and its normal atomic lattice pattern may be different. Such intrusion causes the lattice to be distorted and the alloy is harder and stronger than the parent metal.

(4) When an alloy is made from two metals which possess a markedly different size of atom and which crystallize in different atomic patterns, the range of alloys divides into a number of 'phases'. The occurrence of a new phase is characterized by a change in the atomic arrangement and is also evident when the alloy is examined under the microscope.

(5) When an intermetallic compound (see page 75) is formed, the various kinds of atom present build an entirely new type of lattice, which is sometimes quite complicated and is different from that of either ingredient.

IMPERFECTIONS IN CRYSTALS

From the previous description of the crystalline form of metals, it might be concluded that such crystals are perfectly regular arrangements of atoms, like marbles in a game of Solitaire. However, a great deal of evidence over the past thirty years has confirmed that metal grains or crystals are far from perfect and that many of their attractive mechanical properties are due to imperfections. In particular it leads to an explanation of why most metals are tough and not brittle, why they will stand heavy loads, extend a little and then stop extending; it also helps to explain why metals will endure shock loads and reversals of stress better than most non-metallic materials, which do not have these kind of imperfections.

If one considers the tree-like formations known as dendrites which are illustrated in *Fig. 19*, page 77, it is easy to imagine that a slight bending of each branch while growing from the molten metal leads to a lack of registry when the branches meet and the metal finally freezes. Because the branches are linear, these misfits tend to be in parallel lines between successive branches. Another type of irregularity is a block arrangement known as a sub-grain, about 0·001 mm diameter, where each block of atoms is slightly deranged relative to its adjacent blocks, like a jerry-built brick wall.

The late Sir Lawrence Bragg was the first man to demonstrate the kinds of disarrangements which a regular pattern of uniform atoms might undergo at a metal grain boundary. He displayed this in a simple and rather beautiful experiment, using bubbles on the surface of water. This is the so-called 'bubble-raft experiment'. Plate 6 shows the way in which a large group of bubbles arrange themselves. It might be expected that the configuration of the bubbles would be in absolutely straight lines but the photograph shows a slight change in direction near the bottom right hand corner and a Y-shaped join in the top left hand quarter of the photograph. These illustrate that what can happen with bubbles can happen with atoms, though it must be realized that this is a great simplification of the three-dimensional disarrays which occur in the structure of metals.

The photo-micrographs in Plates 24a, b, c and d show some imperfections or dislocations; furthermore they illustrate how the electron microscope has made it possible to examine metals under magnifications that would have been impossible a generation ago. Plate 24a is an electron microscope picture, at a magnification of 50,000. A piece of aluminium has been given a 10 per cent reduction in thickness. The fuzzy dislocations, like tangled wool, are disposed in a cellular arrangement around sub-grains, the white areas, where there are very few dislocations. When one remembers that normal working operations on such a metal cause reductions of 25 to 50 per cent, it will be realized that such heavily worked metals would show many more dislocations.

Plate 24b, which looks like part of a painting by Miró, shows dislocation lines that start and finish at black spots, which are in fact particles of aluminium oxide entrapped in a sample of brass. The photograph, at a magnification of 10,000, shows how the oxide particles 'trap' the dislocations. The effect is to harden and strengthen the alloys.

Plate 24c shows an iron alloy containing 3·25 per cent silicon, taken at a magnification of 2,200. The three parallel bands are slip bands, which will be discussed again in chapter 9. The dotted line that crosses the slip bands is a dislocation.

Plate 24d is a photo-micrograph of the iron–silicon alloy referred to above. It has been cold rolled to give a 4 per cent reduction in thickness and annealed until recrystallization commences. The dark areas are those where the effect of dislocations remain and the light areas show that, on annealing, new grains have begun to grow.

Although these effects are loosely described as dislocations, there are several associated effects, all of which have a profound influence on the strength of metals:

(1) Point defects, caused by atoms which are small enough to squeeze into the interstices between larger atoms.

(2) 'Vacancies', where atoms are missing.

(3) Line defects, called dislocations.

(4) Surface and inter-face defects, which include grain boundaries and sub-grain or block boundaries.

A knowledge of the inner structure of metals is of more than theoretical interest, for by helping us to understand how metals and alloys are built up, it enables us to exert a more precise and comprehending control over alloy composition and heat-treatment than would otherwise be possible. Also it helps to explain what happens when metals are hardened or stressed. In all branches of science, investigations which at first appeared to be of a merely academic interest have often proved to be of great practical benefit, and the study of the inner structure of metals was no exception to this rule.

8
SHAPING METALS

Consider the making of a sewing needle. A white-hot ingot of
0·8 per cent carbon steel is forced between pairs of rotating,
grooved rolls. This process reduces the thickness and increases
the length of the steel, giving a square-sectioned bar, and is
followed by a further rolling treatment, using grooved rolls of
such a shape that the bar is made into rod of circular section
about 12 mm diameter. In the next stage the metal, when cold, is
drawn through successively smaller holes in hard steel dies which
reduce it in diameter, so that eventually it becomes wire of the
same diameter as the needle. The wire is then finally 'annealed'
to soften it, special precautions being taken that none of the
carbon of the steel is lost by the action of the furnace atmosphere.
The wire is then cut to a length just over twice that of a needle
and each end is pointed by grinding, in a continuous process, on
a rapidly rotating emery wheel. The middle part of the annealed
wire is stamped so that it has the form of two needle heads joined
together and an eye is then pierced in each head (*Fig. 23*).

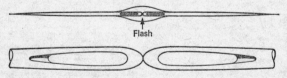

Fig. 23

The twin needle is broken into two and the 'flash' resulting
from the stamping operation is ground away. The pieces of wire
now begin to look much like needles, although they are still so
soft that they can easily be bent double. In the next operation the

steel is 'hardened' by heating the needles to a bright red heat and then quenching. The steel is now extremely hard and brittle, so brittle that a handful of needles can be snapped like macaroni, and before the needle is suitable for use it must be 'tempered', so that whilst maintaining much of the hardness, the brittleness is removed. The tempering involves heating the steel to a temperature of a little over 200°C.

The next process is remarkable. Thousands of needles are placed on a canvas sheet, covered with emery powder and soft soap, and the canvas is rolled up like a roly-poly pudding and revolved between weighted rollers for many hours. In this way the needles are ground by the emery so that they become bright and the points are properly shaped. Different ways of packing the needles into the canvas roll and different times of running result in the variety of point shapes required by users. Many needles are sold in this condition; others are nickel plated and also gold plated round the eye. The needles are now ready for inspection, counting, and automatic packing.

And all this, with overhead and distribution costs, at the price of ten needles and a bodkin for ten pence!

Nearly every metal article has more or less as fascinating a manufacturing history as a sewing needle. Although a multitude of shaping processes are used, they can be grouped into five classes:

(1) Shaping from molten metal
(2) Shaping hot solid metal
(3) Shaping cold metal
(4) Shaping by joining metals
(5) Shaping from powdered metal

The joining of metals is described in chapter 19 and powder metallurgy in chapter 20. The first three classes are discussed below.

SHAPING FROM MOLTEN METAL

For many generations, crucible furnaces have been used to melt metals and alloys. The metal is held in a refractory container which, in the old days, was heated by burning charcoal and later by coke, oil, or gas. Another method is the reverberatory furnace, designed so that the metal is contained in a shallow bath of comparatively large area; hot gas or an oil flame plays on the sloping furnace roof and 'reverberates' heat on to the surface of the metal. For melting non-ferrous metals, reverberatory furnaces hold several hundred kilograms or a few tons of molten metal. The open hearth furnaces for producing steel from cast iron and scrap steel (page 142) are of the same type of construction, though the amount of metal dealt with is very much greater.

In the twentieth century, the use of electricity came into the picture, at first as the electric arc and then using electric resistance elements as the source of heat. As the use of alternating current has been developed, low-frequency electric induction furnaces have been widely exploited. High-frequency electric induction furnaces have become standard equipment for melting certain high-quality alloys.

There have been developments in the technology of handling the newer metals, particularly molybdenum, titanium, zirconium, and some of the 'super-special' steels. Among these developments an important step has been the perfection of consumable-arc–vacuum and protective-atmosphere furnaces. More recently beams of electrons have been used to ensure the highest purity of metal during the melting operation.

There are several ways of producing castings from molten metal; those made from sand moulds are perhaps the best known. Let us assume that a casting is to be made, shaped like that in *Fig. 24a*. A solid pattern or model of the letter M is first prepared and is used to make an M-shaped cavity in moulding sand. The pattern is then removed and the cavity filled with molten metal which solidifies and thus forms a casting.

Moulding sand used in foundries is generally excavated from quarries – there are excellent supplies of naturally bonded sand near Mansfield, Wolverhampton, Erith and Leighton Buzzard. As ex-makers of sand castles will remember, sand of the sea-shore type does not bind well together, because its natural clay bond has been washed out by the salt water; for foundry use sand has to be bonded synthetically with oils, molasses, or fullers'

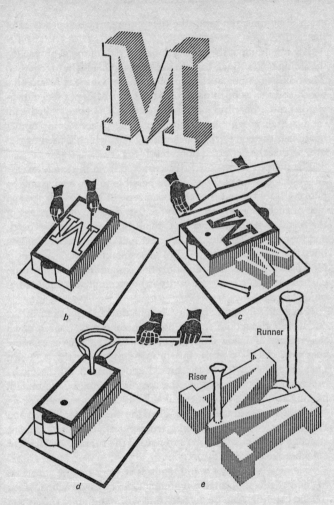

Riser

Runner

Fig. 24

earth. The finer the sand, the better the surface appearance of the casting.

The moulding of the sand is done in a pair of metal or wooden moulding boxes which contain the sand and help it hold together. The boxes are open at the top and bottom. One box is laid on a board and the wooden pattern of the letter M is put face downwards on the board in a central position. The box is filled with sand, which is rammed firmly round the pattern, and another board is placed on top of the moulding box, which is then turned over so that the pattern is uppermost (*Fig. 24b*).

The pattern is now carefully extracted from the sand, leaving an M-shaped cavity. It would be possible to make a casting by pouring liquid metal into this open type of mould, but the top face of such a casting would not be flat, because of shrinkage of the metal as it sets. In practice, moulds are made in two parts, and for casting the letter M the upper half consists of a moulding box, containing sand with a carefully smoothed surface. This is placed on top of the first moulding box containing the impression of the letter M, so that the cavity is now completely enclosed in sand (*Fig. 24c*).

A channel or 'runner' must be cut through the sand in the one half of the mould so that liquid metal can be poured down this runner and flow into the cavity (*Fig. 24d*). One or more channels, called 'risers', must also be made, so that the cast metal fills the cavity in the mould and rises up the riser; the provision of a riser assisting the complete filling of the mould, helping to ensure soundness in the casting, and providing an outlet for the air in the mould. When the metal has solidified and the two halves of the mould separated, the casting remains attached to a neck of metal, representing the runner, which is cut off subsequently (*Fig. 24e*).

Anyone who has not seen foundries at work may be surprised to know that complicated sand moulds can be made which, when treated carefully, can be inverted and will stand up to the stream of liquid metal without being washed away. The choice of the right kind of sand and the strengthening of it with bonding materials makes this possible. Apart from these considerations, the sand must also be packed or rammed to the correct degree.

If it is rammed too tightly, the air contained in the cavity, and water vapour from the heated sand, cannot escape through the sand, but are trapped in the molten metal, causing unsoundness. On the other hand, insufficient ramming may lead to the mould being washed away by the metal when it is cast.

Because of the contraction which occurs on solidification, extra metal is often added or 'fed' after the main bulk has been poured. In casting the huge propellers of the *Queen Elizabeth 2*, described later, the feeding process continued for several hours.

When a casting has to include holes or other complex features, these portions are made by separate 'cores' which are fitted into the two halves of the mould after the pattern has been taken out. These cores are frequently made of oil-bonded sand and are baked to give them strength and rigidity. A complicated large mould may have over a hundred loose core pieces and take many hours to make.

For the production of big quantities of similarly shaped castings, foundries are mechanized, applying the principles of mass-production. The patterns are of metal and the moulding boxes are mounted on a conveyor. The sand, having been mechanically mixed and reconditioned, is automatically flung into the moulding boxes which are vibrated on machines so that the sand is consolidated correctly. The moulds are assembled, and while they are still moving along the conveyor belt the metal is poured into them; after the belt has moved forward and the casting solidified, the mould is automatically tilted and the solid casting removed. The mould box continues and the sand falls out to be reconditioned.

CASTING THE *Q.E.2* PROPELLERS

During the last fifty years great progress has been made in the methods of ship propulsion. Propellers used to be simple shapes made in cast iron or steel in the engine builder's own foundry. Today the design is tested scientifically and the variable pitch propellers, made of high tensile copper alloys, are usually made in specialized foundries with all the facilities for metal melting, mould making, casting, machining and transport required for the propellers of a large vessel.

Four propellers were made for the *Queen Elizabeth 2*, two working and two spare. The alloy chosen was a special nickel–aluminium–bronze containing also about 12 per cent manganese. The cast weight of each propeller was 47 tons; a machined propeller which weighed 32 tons is shown in Plate 9. The material used for moulding was a mixture of pure silica sand and Portland cement. Pits are necessary for moulding large propellers, partly for reasons of safety and partly to lower the top of the mould to a reasonable height for working. A series of massive concentric slotted cast iron rings are bolted into the floor of the pit, to enable the mould to be held rigidly.

The mould for each blade of the propeller was made in two halves. The bed, the upper surface of which defines the pitch face of the blade, and the top, the lower surface of which defines the suction surface of the propeller. The centre line and the approximate shape of the blade were marked out on the moulding site. Wooden shuttering was erected to form a box into which the sand–cement mixture was rammed, together with iron reinforcing bars. The pitch face was then formed by a process known as 'strickling', using a board with a long arm at the one end and a roller at the other. The roller runs on a rail, set so that the surface of the blade form can be generated.

Next the blade pattern was constructed and the top part of the mould made, and left to harden. Each of the six blade forms was made in turn. All the mould surfaces were cleaned, then the top parts of the moulds were placed in position and secured by T-bolts. Steel wire ropes were tightened around the mould. For several hours before casting, hot air was blown into the mould to remove all moisture. The total weight of the mould was 100 tons. The moulding time for each propeller was 710 hours and the mould was dried for 90 hours before pouring began. A mould in the course of manufacture is shown in Plate 8.

The alloy was poured at a temperature of 1,055°C, taking only about 5½ minutes, during which time forty tons of liquid metal was run into the mould. In order to allow for the shrinkage as the metal solidified, the casting was 'fed' after the main bulk had been poured. This process took place in two stages. $3\frac{1}{2}$ hours after the first cast, a further $2\frac{1}{2}$ tons of metal was fed and $2\frac{1}{2}$ hours after that they poured 2 more tons. Both these feeding operations were done at the slightly higher temperature of $1,070°C$.

A few days after pouring, the mould was dismantled and the casting lifted out and trimmed. It was then machined in the bore and on the contours of the blades. The whole propeller was then ground and polished to a very smooth finish. The blade edges were then filed to the designed shape; finally the balance of the propeller was checked and necessary final adjustments made.

Such casting operations as this make one appreciate the motto on the coat of arms of the University of Birmingham Metallurgical Society: 'The hand that wields the ladle rules the world'.

DIECASTING

Anyone who learns about foundry casting may remark that, though the pattern is used repeatedly, it is a pity that the sand mould has to be made over and over again. This handicap is overcome by diecasting, whereby permanent metal dies are used for making large quantities of castings in zinc, aluminium and magnesium alloys. In one simple diecasting process the two halves of the die are made of steel or cast iron, but the general operation is similar to that of foundry casting. The metal is poured and the die is opened and closed manually. This process is called 'gravity diecasting', because the metal enters the die under its own weight.

At the beginning of the twentieth century a more advanced process, pressure diecasting,* was developed, and today it is one of the most versatile ways of mass-producing small and medium-sized castings of great accuracy and good surface appearance. In pressure diecasting the operation of the die and cores is semi-automatic or, in some cases, fully automatic. The liquid alloy is forced into the die under pressure, ranging from less than 100 kg to over 1,500 kg per sq. cm, depending on the type and size of the machine. The pressure of the injection causes a precise reproduction of the form of the die; a dimensional accuracy of plus or

*In the U.S.A., gravity diecasting is known as permanent mold casting; pressure diecasting is known as die casting.

minus 0·05 mm can be obtained, with even closer limits in some cases.

Diecasting machines are divided into two main types. For alloys of rather low melting point, and especially for zinc alloys (see page 218), the molten alloy is held in a container which is part of the machine. A plunger, permanently immersed in the molten metal, is forced downwards under pneumatic or hydraulic pressure and this causes a 'shot' of alloy to be injected into the die cavity. Such a machine is called a hot-chamber diecasting machine.

The other type, known as a cold-chamber machine, is generally used for pressure diecasting alloys of aluminium, magnesium, or copper, all of which have higher melting points than that of zinc alloy. The molten alloy is held in a crucible or other melting unit adjacent to the machine. An amount sufficient for one shot is ladled, either manually or automatically, into the plunger cylinder of the diecasting machine. A hydraulic ram then forces the metal into the die cavity.

When the diecasting has been cast, it cools very rapidly; the die is provided with a number of separate water-cooling channels, so that each portion of the die is cooled at an appropriate rate and the diecasting solidifies uniformly. The die is then opened, the half furthest from the 'sprue', where the metal entered, moving away from the stationary half. The design of the casting is arranged so that it adheres to the moving half of the die. It is forced away from this face by ejectors, which are cylindrical steel rods 3–10 mm in diameter, disposed in suitable positions in the die block, around the form of the diecasting. The efficiency and maintenance of ejector pins play an important part in the economic production of diecastings. To obtain greater reliability, there has been a trend to increase the diameter; sometimes ejectors of as much as 25 mm diameter are used. The ejectors push the diecasting away from the die face so that it can be removed, either by tongs or by arranging that it falls into a bath of water, from which a conveyor takes the diecasting to the next operation.

The development of pressure diecasting has been fraught with many technical and metallurgical problems. Speed of operation is required, so the process has been more and more mechanized. Some small components, for example, zip-fastener elements, are automatically diecast, at the speed of 400 per minute, one maintenance engineer alone supervising the operation of a battery of several machines. Such diecastings are made on hot-chamber machines and it has been comparatively easy to make this type operate automatically, especially for small items. The development of automatic cold-chamber machines, for aluminium alloys, has been more difficult but is now well within the bounds of possibility. It is necessary to arrange that the molten aluminium is transferred automatically from the crucible to the machine; this is sometimes done by a mechanical ladling device. In another process the aluminium is in an electrically heated container; air pressure thrusts the correct amount of molten alloy through a sleeve into the cylinder of the diecasting machine.

Mechanized operation leads to increased speed of output, but it is necessary to control the cycle of operations so that the diecasting cools rapidly and uniformly; if the die began to open before the castings were fully solidified, a defective product would be made, and the cycle of operations would be held up. If the diecast metal did not fill the die properly, porosity cavities would be formed. As in many metallurgical processes, mathematics is becoming a necessary skill for the technician; the factors of injection pressure and speed, the size of the opening through which the molten metal is injected into the die, and the amount of water cooling which must be applied, are all interrelated; thus the design of the diecasting die, which only a few years ago was a matter of rule of thumb, now depends on the understanding of quite complex applied mathematics.

For thirty years or more gravity diecasting has been used, not only for medium-sized castings, but for large aluminium alloy components weighing up to 300 kg. So far pressure diecastings have not reached this magnitude but the diecast automobile cylinder block, discussed on page 196, indicates the shape of things to come. A cylinder block die weighs as much as 30 tonnes, the machine is over 15 metres in length and is a size which dwarfs the operator. At the moment of injection of the molten metal the die halves must be held firmly together; immense locking forces are required for cylinder blocks and components of similar size. Plate 10a illustrates one of the largest diecasting machines in Europe; it has a locking force of 2,500 tonnes. The metal will be injected from the left; the die halves will be bolted on the two platens shown to the left of the two supervisors; the links which operate the moving platen are on the right.

The use of vacuum is extending into diecasting, the air in the mould cavity being withdrawn before the diecast metal is injected. Under normal conditions the incoming molten metal pushes before it the air previously contained in the die cavity. Although good design and the provision of minute gaps, known as vents, between the working parts of the die, usually allow most of the air to escape, there are some complex diecastings where this is not completely possible and where there is a risk of air being entangled in the injected metal, thus causing porosity. The prior removal of air from the die cavity by vacuum extraction makes it possible to produce sound castings and to reduce the percentage of scrap that might otherwise be produced if vacuum were not applied.

Vacuum diecasting is one of the many new processes which may in the future help to produce metal products of improved quality and performance. So far it is hardly past the experimental stage, but, as experience grows, vacuum diecasting methods may be used to quite a large extent.

CENTRIFUGAL CASTING

For casting pipes and similar shapes, a permanent metallic cylindrical mould, without any cores, is spun at high speed and liquid metal poured into it, so that centrifugal force flings the metal to the face of the mould, thus producing a cast hollow cylinder of uniform wall thickness. This process yields a product having a dense, uniform outer surface; consequently a drain pipe or cylinder liner cast by this method is considered superior to similar ones cast in sand moulds. Cast-iron piston rings are cut from such cylindrical shapes made by centrifugal casting, and the process is now being used to make complicated components.

INVESTMENT CASTING

Originally investment casting was known as the 'lost wax' process and was confined to statuary, art and jewellery. In modern times the process has been adapted for engineering and functional uses. It is a process for producing close-tolerance castings by an expendable pattern technique. With normal casting, the mould must be in at least two parts to enable the pattern to be removed prior to casting in the case of expendable moulds, or for the casting to be removed from a permanent mould. The differences brought about by the investment technique are the elimination of the two-part mould and the use of pre-formed cores for the production of internal cavities in the casting.

In simplest terms a pattern is formed in wax or a plastic material, or by freezing mercury. This pattern shape is covered with a slurry of a refractory material which is compacted and solidified so that, when the pattern material is melted out or dissolved away, a mould cavity is left into which molten metal can be poured. Clusters of such individual moulds can be arranged to feed from a central distribution spout or 'down gate' to speed production.

The advantages of the process are that no parting line is seen, undercut sections are easily formed, no taper is required on the casting, the complexity of shape presents few problems and a very smooth surface and accurate dimensions are obtained. Most high melting point alloys and expensive metals such as gold and silver are readily cast in this way. Gold rings and a large proportion of costume and exquisite jewellery are investment cast, coupled with centrifuging during the casting process. Aluminium and copper alloys, stainless steels and the exotic alloys used in jet engines are investment cast.

CONTINUOUS CASTING

Almost all metals and alloys must be cast into a suitable shape prior to working. Rectangular slabs are the first stage for flat products such as strip and sheet; square sectional long billets for rolling into rod and wire; cylindrical shapes for extruding into rod or tube.

Until the mid 1930s all such shapes were invariably made by casting into a metal mould having a cavity of the required form. About that time the need arose for producing aluminium alloys in a fine crystalline form in the cast condition. This requirement led to the development of continuous casting. A water-cooled copper or aluminium mould about 150–250 mm deep, having the shape of the desired cross-section of billet or slab shape is sealed at the bottom by a retractable base plate (*Fig. 25*). Molten metal is poured into the cavity continuously while the base plate is slowly lowered; the metal solidifies first as a shell adjacent to the mould and solidification progresses into the centre of the cavity. The process is hastened by additional cooling, by water jets or sprays and by withdrawing the billet into a tank of water below the mould. Almost all aluminium alloys are now semi-continuously cast; a length about 3–6 metres is poured, casting is stopped, and the solid metal withdrawn. Casting is recommenced by returning the base to its original starting position.

This development was relatively easy for metals melting at up to 700°C, such as aluminium or magnesium, but it is only in the last fifteen years that the metallurgical and economic advantages of continuous casting of iron and copper base alloys have been

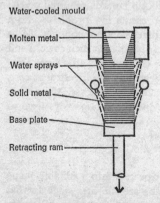

Water-cooled mould

Molten metal

Water sprays

Solid metal

Base plate

Retracting ram

Fig. 25. Semi-continuous casting of billets or slabs

realized. For such large tonnages it may be more economic to cast the metal continuously and cut the solid metal off well below the mould by means of a moving saw as shown in *Fig. 26*. The speed of descent of the solid casting is limited according to the thermal conductivity and the casting characteristics of the alloy or metal; thus for aluminium alloys speeds of 70–400 mm per minute are practicable. The continuous casting of steel is discussed on page 151 and illustrated on Plates 11 and 12.

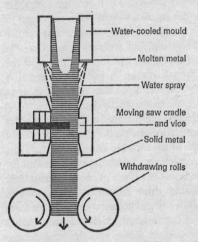

Fig. 26. Continuous casting of billets or slabs

THE SHAPING OF SOLID METAL

If a bar of cold steel is hammered, a great amount of energy is needed to change its shape permanently. The same metal, when heated to bright redness, is soft and pliable; hammer blows will alter its shape easily. The village blacksmith makes use of this and with his hammer, anvil and forge shapes hot steel into horseshoes or repairs tractors.

With the coming of the machine age, attempts were made to develop mechanical hammers which would forge larger shapes than even the mighty smith was capable of tackling. One old machine is the 'tilt hammer', which can be seen at the Abbeydale Industrial Hamlet in Sheffield. An iron hammer head, weighing about 50 kilograms, is fixed at the end of a long wooden arm, pivoted at the centre. A cam rotates beneath the arm, so that the head rises and then drops by its own weight on to the piece of hot metal held beneath it.

A modern development of the same principle is the drop hammer, where a heavy steel die block, working between two vertical guides, is mechanically lifted about $1\frac{1}{2}$ metres above the anvil and allowed to fall under its own weight on to the metal to be forged. Two halves of a die are made, by a somewhat similar method to that used in diecasting, but without cores. One die block is bolted to the anvil at the base of the machine and the other is fixed on to the drop hammer. A hot bar of metal is firmly held on the anvil by tongs and the hammer falls so as to forge the metal between the two halves of the die.

For a fairly complicated part, the die may contain three pairs of impressions of progressively increasing detail, the purpose of which is to change the shape of the metal in gradual stages. Thus three forging operations are performed with the same die and for each of these the workman holds the piece of metal successively in the three positions during three successive blows of the hammer. The first impression causes the metal to assume a form roughly approaching that of the finished article, the next brings the metal practically to the desired shape, and the final die impression makes the forging accurate in dimensions.

A modification of drop-forging employs mechanical or steam power to push the hammer downwards, thus increasing the power of the blow. In such methods, and also in drop-forging, the hammer blow is a rapid one and certain disadvantages are connected with the sudden action of the hammer. If, for example, the barrel of a big gun were forged in this way, the outside of the metal would receive the blow, but the effect would not be completely transmitted to the interior. Now one valuable feature of forging is that it improves the strength of the metal by refining the structure and making it uniform, so for heavy forgings, such as marine propeller shafts, a different method is adopted to work the metal throughout. This consists of gradually squeezing the metal by pressure in an immensely powerful hydraulic press. The hammer or 'ram' is pressed downwards under a force sometimes as great as 10,000 tons.

Although the hydraulic forging press is a more expensive equipment than a drop-forge, it has advantages besides that of giving greater strength to large components. On account of the much higher pressure that is used, it operates with less noise and vibration than the drop-forge.

EXTRUSION

The process of extrusion involves a similar principle to that of squeezing tooth paste from the tube or making cake decorations by squirting icing sugar from a bag. A prodigious pressure would be necessary to extrude most of the common metals while cold, and the plant is generally designed to extrude the metals hot, for example, at a cherry-red heat (700–800°C) for copper and brass or at between 400° and 500°C for aluminium alloys.

An extrusion press of average size will take a piece of cast brass in the form of a cylinder about 120 mm diameter and 700 mm long, weighing about 60 kg. This is heated to 750°C in a furnace and then placed in line with the container of the press. A ram pushes the metal into the press and, under a high pressure, a rod of hot metal emerges through a die at the end of the container. Thus, within the space of fifteen seconds a block of metal is extruded into a rod 30 mm diameter and 10 metres long.

The principle of the type of machine described above is illustrated in *Fig. 27a*; the process is known as 'direct' extrusion. Another method, known as the 'indirect' process, is illustrated in *Fig. 27b*. This has some technical advantages, such as less power required and greater uniformity of structure of the metal, but it involves a more intricate type of machine. The main difference between the two methods is that in the direct process the metal is forced through the die, while in the indirect process the die is forced through the metal.

Die orifices are made in a variety of different shapes so as to produce different sections such as curtain rails, windscreen and window sections for automobiles, and even gear wheels for small

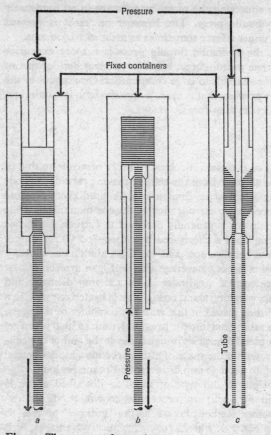

Fig. 27. Three types of extrusion

machinery, in which case the continuous gear-shaped length is first extruded and is later sectioned into gears. The process is so economical that for copper, aluminium, magnesium, and lead alloys extrusion is a normal production procedure for such products as rods, bars, tubes, and strips. The extruded rod or section is sometimes finally drawn through another die in the cold state, in much the same way as for wire drawing (*Fig. 31*). This is in order to improve the dimensional accuracy and in some cases to increase the strength of the alloy.

Extrusion forms a useful starting-point for the production of tubes and pipes, since by a modification of the press it is possible to arrange for the metal to be squeezed between a solid steel mandrel and a die as shown in *Fig. 27c*. Plumbers' lead pipe is made in this way, while as an example of the adaptability of the process one may mention the lead tubing which is extruded and shrunk directly on to finished cotton-covered insulated wire, thus forming lead-covered cable for underground telephone communication.

THE ROLLING OF METALS

Although as early as the sixteenth century soft metals such as gold and silver were rolled into sheets it was not until the eighteenth century that the process began to be used industrially, owing to the work of an Englishman, Henry Cort. The principle is illustrated in *Fig. 28*, which might be compared with the picture of a large rolling mill for steel shown in Plate 10b.

Metals may be rolled either hot or cold. The advantage of

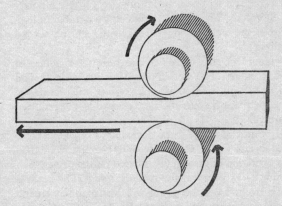

Fig. 28

working a metal hot is that it can be reduced in thickness more easily than when cold rolled. On the other hand the surface condition and accuracy are not so good as those obtainable by cold rolling. The most general use of hot rolling, therefore, is for 'breaking down' large ingots; cold rolling is used to make smooth and accurate thin sheets of metal.

The end products produced by rolling are so diverse that it is difficult to give a general description but a large works might produce rods in the following way. Ingots of steel weighing about 8 tons come from the adjacent steel works; they are put in a 'soaking pit' furnace to bring them to an intense white heat and to ensure that the temperature of the metal is uniform. At the next stage a pair of heavy rolls, driven by a 6,000 h.p. motor in a 'cogging mill', have a number of grooves and as the steel is passed backwards and forwards several times through the rolls it travels through a different pair of grooves at each pass. A set of rolls is called a stand; often there are several stands, forming a mill train.

The rolled ingot, now in the form of a long thick bar, is reheated and further reduced in a roughing mill and then in a finishing mill. These latter two are operated by one motor, of about 4,500 h.p., and are end to end, so that the one power unit can run them both; it will be noticed, however, that the power required is enormous. The finished product is a number of long lengths of rod, the steel having been sheared during several stages of rolling.

A great amount of steel is rolled into beams for industrial buildings and bridges. The well-known I-section is first rolled to a 'dog bone' shape; a subsequent set of four rolls forms the shape of the web and flattens the flanges so that the required section is formed.

Universal beam mills have been developed to make column sections and beams more accurately and competitively than before. The operation of these mills is based on the principle of four-roll contact – two vertical and two horizontal. The beam is passed to and fro between these rolls which shape the web and flange. A separate stand with horizontal rolls works the edges of the flanges. The two stands work in tandem, doing both the roughing and finishing rolling on an ingot of steel which has previously been shaped on a cogging mill.

Other high speed mills roll steel into rods for nails, screws or wire for fencing and suspension bridges. Enormous lengths are produced from many stands of rolls, sometimes up to twenty in number, with their rolling speeds so matched that the rod zips through the final stages with the speed of a bullet.

Because of the very large amounts of steel sheets required for motor car bodies, metal cans, galvanized sheet and many other requirements, rolling mills to make sheet in bulk have been developed to a high degree of mechanization. White-hot ingots of steel weighing 25 to 50 tons are first flattened in large reversing mills, to make slabs from 125 to 175 cm wide. Then the slabs, while still hot and 10 cm thick, are rolled in a series of four-high mills, each driven by motors of about 3,000 h.p. These successive stands of rolls are usually arranged in twos or threes; the slab is passed through, continuously reducing the thickness. The enormous length of strip, now as little as 5 mm thick, is immediately coiled hot, at speeds of 300 to 700 metres per minute.

These hot-rolled coils are then reduced further by cold rolling. Most cold-rolling mills producing steel strip and sheet are known as tandem mills, which can be 2, 3, 4 or 5 stands, at about 4 metres apart. The rolls are made of forged steel. In order to provide the required accuracy in the finished product the working rolls are backed up top and bottom by other rolls at least three times the diameter of the work rolls; their function is to support the work rolls, preventing them from distorting. At each stand of rolls the gap is progressively less, so that, as the strip of metal goes forward, its thickness becomes less and its length increases. As the metal is progressively reduced in thickness the speed of the rolls must be accurately controlled to avoid buckling or breaking of the steel. This is done by small 'tension rolls' over which the strip passes between stands. If there is any tendency to buckling, the tension roll records the drop in pressure and the setting of the work rolls is rapidly corrected. Since the metal issues from the rolls at increasing speed, finishing like an express train, the safety controls must be precise. The loading or squeezing power of the rolls, the adjustments for controlling the thickness of the strip, the lubrication, the mechanism which directs the metal from one stand or mill train to the next, must be controlled automatically. The rolls, their housings and bearings, are designed so that in case of breakdown rapid replacement is possible. The work rolls, weighing only a few tons, are changed regularly, requiring only about 15 minutes. Occasionally, when the backing rolls and housings need changing, this involves moving over 50 tons.

The power required for cold rolling is greater than that for hot and the requirements of accuracy and surface finish are more precise. One new cold rolling mill in U.S.A. accepts strip up to 5 mm thick and from 100 to 180 mm wide, weighing up to 40 tons, and cold-rolls it down to 0·15 mm thick. The mill has five sets of rolls in tandem and as the strip passes from one to the other it gathers speed till the last set of rolls is delivering sheet at 1,500 metres per minute. All the sequences of operations, the tracking of the coil of metal, and the regulation of the rolls are computer controlled, and closed circuit TV is used to assist supervision.

As cold steel is harder than hot steel, cold rolling mills need harder rolls and, as mentioned above, the work rolls have to be backed up to prevent distortion. In the Sendzimir cold mill the work rolls are driven by friction from the strip, which is pulled through the roll-gap by tension of the coiler mechanism. The work rolls are backed up by a cluster of heavier rolls. The general principle of the Sendzimir mill is illustrated in *Fig. 29*. This type of mill is in wide use, and especially for rolling stainless steel. One advantage is that roll changes take only a few seconds since the small, hard work-rolls can be withdrawn by hand from this position and replaced by newly reground rolls. As in all cold-rolling mills the Sendzimir rolls are flooded with a coolant while in operation, so as to prevent heat from the tremendous surface pressure leading to expansion of the steel on its way through.

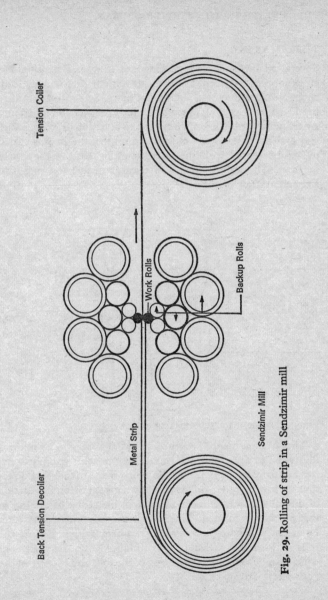

Tension Coiler

Work Rolls

Backup Rolls

Metal Strip

Sendzimir Mill

Back Tension Decoiler

Fig. 29. Rolling of strip in a Sendzimir mill

TUBE-MAKING

Several methods of tube-making are in use. Extrusion has already been described, representing a method of making lead, copper, magnesium, and aluminium tubes. Some steel tubes are produced by a welding process in which the metal is rolled into a narrow strip which is then heated to a very high temperature and drawn through a die of such a form that the strip is curled to a tube shape, with the edges abutting so that they weld together. The welded tube is afterwards treated to bring it to size and to straighten it.

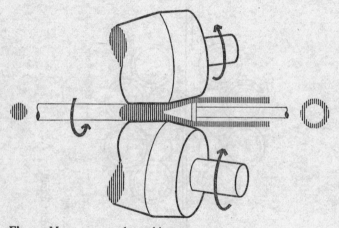

Fig. 30. Mannesmann tube-making process

The Mannesmann process for making seamless tubes is illustrated in *Fig. 30*. A solid roll of hot metal is spun between two mutually inclined heavy rolls which rotate in the same direction so that the rod is pulled forward between them. The action of the rolls draws the metal away from the centre of the rod so that a cavity is formed. This hole is shaped to a circular form by means of a nose-piece placed on a mandrel just beyond the rolls and a thick walled tube is produced, the dimensions of which can be varied by the setting of the rolls and the size of the nose-piece of the mandrel.

The processes so far described are for producing tubes from hot metal. If a tube of good surface-appearance and strength is required for aircraft, hypodermic needles, steel furniture, or bicycles, a further process of cold drawing is applied in which the tube, pointed at one end, is pulled through a hole in a die of slightly smaller size than the original diameter of the tube, which is reduced in diameter accordingly.

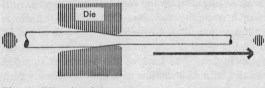

Fig. 31. Wire-drawing

THE DRAWING OF WIRES AND RODS

Wires are produced as shown in *Fig. 31* and are always made at a comparatively low working temperature. The raw material for making wire is hot-rolled rod about 6 mm diameter. After annealing, cooling, and removal of scale, the steel rod is pointed at one end and then inserted through a hole in the die slightly smaller than the size of the rod. The pointed end is gripped from the other side and the rod pulled through, thus reducing its diameter. In modern continuous methods, particularly applied to copper, the wire is drawn through a first die, turned round a roller, then passed through a second die, and the same procedure is repeated eight or nine times, using successively smaller holes and rollers of increasing peripheral speed. By the time the wire enters the last die it is much reduced in diameter and is travelling at high speed. In order to save continual re-threading of the machine, lengths of rod are welded together before drawing.

'Silver steel' is made by a similar process to wire-drawing, a rod of steel containing about 1 per cent of carbon being cold-drawn to reduce its diameter. The steel becomes still harder by this process and afterwards it is subjected to a very accurate grinding and polishing process. Silver steel contains no silver;

it is the appearance of the metal which gives it this familiar name.

OTHER METHODS OF SHAPING COLD METALS

A button such as is fitted on uniforms is made from metal strip in a number of stages; circular blanks are punched from a strip of annealed brass or nickel silver and each blank is placed in turn on the anvil of a stamping machine. One half of the die is mounted on a heavy steel block which is raised to a given height and dropped on the metal blank, thus stamping the pattern on the front of the button. The back of the button is stamped separately into the form of a shallow cup, and the shank which forms the eye is inserted. In shaping both the back and front a lip is made at the edges and the button is completed by forcing the lip of the back of the button underneath that of the front.

Other metal parts are shaped by pressing in great hydraulic presses. Thus most motor-car bodies are made in this way, the entire roof or back of the car being pressed from a sheet of mild steel in one operation.

Collapsible tooth-paste tubes, radio-coil cans, and patent medicine containers are made by the process of 'impact extrusion'. This method, akin to the extrusion process described on page 103, depends on punching a small blank of cold metal in a die. The metal is squeezed between the die wall and the punch, producing a hollow, thin-walled container. Some aluminium teapots and hot water bottles are made from sheet metal by 'spinning'; nails are made from wire by a continuous cold-heading and pointing operation; cartridges are made by 'deep drawing' (see page 205).

The minting of coins is an important, though somewhat exclusive, branch of the shaping of metals. In 1969, when the Royal Mint was making the old British coins, supplying overseas orders, and preparing for 1971 decimalization, the then record number of 2,685 million coins was made, comprising 1,288 million at the old premises at Tower Hill, 1,382 million at the new works at Llantrisant and 15 million at a training factory at Bridgend. In that year 4,700 tons of cupro-nickel and 4,000 tons of bronze were used for British coinage. In 1970 a still greater number of coins was produced, amounting to 2,796 million.

The Annual Report of the Royal Mint, always an interesting and sometimes a witty publication, gave some information about the die life for coinage in 1969. Although all the dies are made of the same heat-treated high-carbon steel, the shape and size of the coin has a pronounced effect on the life of the die. Furthermore the skill with which the coinage die is used affects the die life. *Table 8* shows the average number of coins per pair of dies, first for the production works at Llantrisant and then at the training works at Bridgend which was operated during the period that production was being moved from London to South Wales.

Table 8

Denomination	Average number of coins struck per pair of dies	
	Llantrisant	Bridgend
10p	65,980	—
2p	128,300	27,655
1p	91,180	19,971
½p	94,770	17,072

Improvements in coinage techniques resulted in a marked improvement in die life. For example at Llantrisant the life of the 2p dies rose from 88,000 coins per pair at the beginning of 1969 to 146,000 at the end.

We have left to the last the cutting, or machining of metals. All metals can be cut but there are great differences in their resistance to cutting; for example it is much easier to cut an aluminium milk bottle top than a thin razor blade with a pair of scissors. It does not follow that a soft metal can be machined easily at high speed, say by turning in a lathe. The soft metal being cut may build up on the cutting edge of the tool and interfere with further cutting. Some metals when machined cause rapid wear of the tip of the cutting tool, because of an abrasive effect; for example some aluminium alloys contain included aluminium oxide which causes rapid tool blunting.

To ensure high output speeds and rapid machining many alloys are made with added elements which make fast machining possible. Small, rounded, low melting point constituents evenly distributed through the metal assist rapid machining. Well-known examples are steels containing lead, or manganese sulphide particles; copper and brasses containing lead, selenium or tellurium particles.

Ever since metals were first employed man has used his ingenuity to devise new ways of shaping metals and there are still no signs that his technical resourcefulness in this respect is being exhausted.

9
TESTING METALS

Metals are tough; they are employed where high stresses and strains have to be endured. A rod of steel 30 mm in diameter – just over an inch – can support a load of over 25 tonnes without fracture. *Fig. 32* attempts to convey some idea of the load which could be borne by such a slender piece of steel.

Different kinds of stress may be experienced in service; for example, the hauling rope, made of stranded steel wire, supporting the cage in a mine shaft is subjected to pulling, or 'tensile' stress, while the vertical columns supporting a bridge suffer mainly compressive stress. Almost all moving parts of machinery undergo rapid changes or combinations of stress; for example, the connecting rod of a steam engine is subjected to alternations of tensile and compressive stresses, while the axle of a railway carriage suffers a combination of bending and twisting.

One of the tasks of the metallurgist is to specify suitable alloys which will endure the stresses encountered in service, and metals must be tested so that their mechanical properties, and especially their strength properties, can be assessed and compared. When a metal is selected for service, mechanical tests must be carried out as an inspection routine in order to make certain that, throughout the production of batches of that metal, the quality is maintained. The principal method is to test to destruction representative samples of the metal, thus subjecting them to much more severe conditions than they will normally endure. The designer estimates the likely stresses in the part he is designing and, with a knowledge of the mechanical properties of the metal, he is able to calculate the shape and size required, allowing a factor of safety.

The most usual mechanical test, the tensile test, determines the force required to stretch and break the metal. Before describing this, it is necessary to discuss a new standard for the measurement of force which, though logical, may at first cause some difficulties before it comes into general acceptance.

Fig. 32. The strength of steel

THE NEWTON

Tensile strength is measured in force per unit area; engineers in Britain have usually stated the strength of metals as 'tons force per square inch', abbreviated in practice to 'tons per sq. in.'. In the U.S.A., tensile strengths are generally stated as 'pounds (force) per square inch'. This particular convention, although widely used, can lead to confusion of thought between the mass unit, the pound, and the pound force (which is the *force* exerted on a pound *mass* by gravity).

A similar situation exists with the metric technical system of units widely used on the Continent. In this system the mass unit is the kilogram (kg) and the force unit the kilogram force (kgf) and here again the distinction is often blurred in practice. Neither of these systems is 'coherent', a coherent system being defined as one in which the product of any two unit quantities is the unit of the resultant quantity.

With the development of metrication in Britain the opportunity was taken to adopt the 'Système International d'Unités', for which the abbreviation is 'SI' in all languages. This system is gaining acceptance and is being legally adopted by all major European countries and in many other parts of the world. In SI, the mass unit is the kilogram and the force unit has a distinctive name, the newton, commemorating the great scientist Sir Isaac Newton and his work on the force of gravity. The newton (N) is the force required to produce unit acceleration on a mass of one kilogram.

From the well-known equation: Force = Mass × Acceleration, it can be seen that the force required to produce standard acceleration (9·807 metres per second) on a kilogram mass is

$$\text{Force} = \text{Mass (1 kg)} \times \text{Acceleration (9·807)}$$

The force of 9·807 newtons is equal to the kilogram force in the metric system. A tensile strength previously expressed as 1 ton (force) per square inch becomes, in SI, 15·44 newtons per square millimetre. The derivation of this conversion factor can be seen from the following:

$$\frac{1 \text{ ton (force)}}{1 \text{ (inch)}^2} = \frac{1{,}016 \text{ kilogram (force)}}{(25·4 \text{ mm})^2} = \frac{1{,}016 \times 9·807}{645·16}$$

Therefore one ton force per square inch equals 15·44 newtons per sq. mm. When kilograms per square millimetre are converted to newtons, the factor 9·807 is used.

In *Table 10* on page 134 we have shown mechanical strengths of various metals in tons force per sq. in. and newtons per sq. mm. A similar comparison has been included in *Table 7* on page 84.

THE TENSILE TEST

The general principles of measuring the strength of metals are similar, but since *Fig. 32* has pictured, somewhat unscientifically, the strength of mild steel, the more conventional method as used in testing laboratories will be described. A typical test piece would be taken from a rod of mild steel, about 130 mm long, accurately turned on a lathe so that its diameter across the narrow parallel part of the section is 13·82 mm; its cross-sectional area is therefore exactly 150 sq. mm. The shape of the finished test piece is illustrated in *Fig. 33*. Two marks are punched or scribed on the parallel portion, at a distance of 69 mm apart; they will be used in measuring the amount of stretch which the metal undergoes. The test piece is placed in a tensile testing machine (Plate 7), which is capable of applying a steadily increasing and measured pulling force on the test piece and thus recording the tensile strength of the metal.

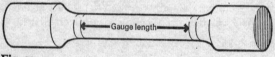

Fig. 33

The stress is gradually applied, but though the recording dial soon shows that a 3,000 kg force is being exerted the steel appears unchanged. If, however, the distance apart of the two marks is measured with an accurate instrument, while the steel is still under tension, it is found to have increased slightly, less than a quarter of a millimetre. If now the force of 3,000 kg is removed, the steel returns to its original length. In other words the metal has so far behaved elastically.

When the tensile force reaches about 4,000 kg an important stage is reached in the process of stressing the piece of metal. This stage is known as the 'elastic limit' and indicates the maximum dead loading to which the steel can be subjected without deforming permanently. If the load were increased to over 4,000 kg force and again removed, the steel would not return to its original length but would remain permanently stretched. With this test we are describing the elastic limit that was reached at 4,000 kg force applied to a piece of steel whose cross-sectional area is 150 sq. mm; the stress therefore is about 27 kg force per sq. mm at the elastic limit. In the elastic range, the metal, when stressed in tension, stretches only a very small amount, but after that an increase of stress causes visible extension which can be measured with a pair of dividers. At above 8,000 kg the steel continues stretching and as the load further increases a neck gradually becomes apparent at the centre, until finally the test piece snaps at the necked portion, the broken pieces being shown in *Fig. 34.*

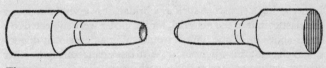

Fig. 34

The original cross-section area of the test piece was 150 sq. mm, and this area is used in the final calculation, which shows that the steel broke at about 55 kg force per sq. mm. Converting now to newtons as shown on page 117 the tensile strength was 55 × 9·807, or about 540 newtons per sq. mm. Had the same test been done in the 'pre-SI-unit' era, its tensile strength would have been stated as about 35 tons per sq. in.

The two broken pieces of the best bar are then fitted together and the distance measured between the two marks which were originally 69 mm apart. Owing to the stretching of the metal before the break, these marks are now found to be 87 mm apart, representing an elongation of 26 per cent. This figure is recorded as a useful guide to the ductility of the steel.

Recently a range of sizes of test bars has been developed for small, medium or large pieces. The measurements of the test bars increase proportionally to the cross-sectional area. For example a gauge length of 80 mm is accompanied by a diameter of 15·96 mm, which gives a cross-sectional area of 200 sq. mm. A gauge length of 56 mm has a diameter of 11·28 mm and a cross-sectional area of 100 sq. mm.

In the test which has just been described, the stressing and eventual breakage of the steel proceeded in two stages; first there was an *elastic* range in which the metal did not distort permanently under stress, then there followed a *plastic* range at higher stresses, in which the metal underwent permanent distortion. Three properties have been determined, the elastic limit, the tensile strength, and the elongation; this information can be put to practical use by an engineer or designer, as three simple illustrations may show.

The Elastic Limit

In designing a structure such as a bridge, it is essential to know the elastic limit of the material, for if a girder were subjected to a stress above its elastic limit it would suffer a permanent dimensional change; this might increase stresses dangerously in other girders connected to it, and lead perhaps to collapse of the whole structure.

The Tensile Strength

A ship's hawser might be subjected to an unusually severe stress, for example when towing a crippled ship in a gale, and the tensile strength indicates the greatest stress that could be applied without the metal breaking.

The Elongation Figure

If a metal is intended to be shaped, for example, by a deep drawing operation, it is essential that it should be ductile; in other words, its elongation figure should be high (the figure given for steel we discussed on page 119 would be counted as moderately good). A metal with only a low elongation would crack when subjected to deep drawing and would be unsuitable for making, say, the body of a fire extinguisher or a shell case, where the metal used is subjected to considerable deformation during shaping.

OTHER MECHANICAL TESTS

During recent years it has become customary to measure one other figure, namely, the 'proof stress', which is the amount of stress required to cause a permanent stretch, usually selected as 0·069 mm on a 69 mm length. Such a figure is known as the 0·1 per cent proof stress; in some metals it is an equivalent of the elastic limits and is easier to measure. Since 1971 there has been an increasing tendency to specify an 0·2 per cent proof stress, especially for non-ferrous metals. However, as this has not yet completely replaced the older measure, any proof stress figures given in this edition are 0·1 per cent.

Hardness is also usually determined during the testing of metal products. Although it is such a familiar word, the meaning of hardness is difficult to define with precision, but technically it may be taken to mean resistance to deformation, and this is the basis of the usual hardness tests. A hardened steel ball or a diamond point is pressed into the prepared surface of the metal for a given time under a given load. If the metal is soft a large indentation is made, while if hard the impression is small; the area of the indentation is determined and the hardness figure calculated on the basis of load supported per unit area of the indentation. Dr Johan August Brinell, of Sweden, devised the best known method of hardness testing, and the Brinell hardness number is frequently used in comparing the hardness of different metals and alloys. More recent forms of hardness testers, based on similar principles, are the Rockwell machine, which was devised in America, and the Vickers diamond-pyramid hardness-tester, designed in England; in both machines a diamond is used to make the indentation. The hard diamond does not deform under load, whereas a steel ball used for testing very hard metals would itself deform and the hardness figure in such a case would be inaccurate. The indentation made on these machines is also rather easier to measure and the hardness of thinner sheets of metal can be measured more accurately than by the ball method.

The Brinell hardness* of some well-known metals and alloys, together with their tensile strength, proof stress, and elongation figures are given in *Table 10* on page 134.

Other methods have been devised from time to time to assess the mechanical properties of metals. Many of these are simple workshop tests; for example, a strip of metal may be tested by counting the number of times it can be bent backwards and forwards over a radius until it breaks. Another more elaborate test on sheet metal consists of forcing a hemispherical plunger into a clamped piece of sheet until it just fractures; the depth of the hollow thus formed is then measured. This is known as the Erichsen test.

In the testing of that sophisticated metal structure the Severn Suspension Bridge a simple 'wrap test' was used to verify that the galvanized wire used in the cables was satisfactory. A specimen of the wire had to be wrapped twice around a mandrel of diameter equal to three times the diameter of the wire. The steel had not to fracture nor must there be any evidence of flaking or cracking in the zinc coating.

Under conditions of very sudden loading or shock, some metals behave differently from what one would expect on the basis of the tensile test alone; for example, steels of certain grades possess high tensile strength but are weak under impact. On the other hand, another type of steel with a lower tensile strength might withstand a severe impact without failure. One type of testing machine, much used in Britain, was developed by Edwin G. Izod to test the behaviour of metals under impact. A notched metal test piece is broken by a heavy swinging pendulum and the amount of energy required to break it is measured.

NON-DESTRUCTIVE TESTING

In most methods of mechanical testing, part of the metal sample is fractured or damaged; one can never be absolutely certain the test-piece is in every way typical of the bulk of the metal to

---

*Various machines use different scales of hardness but these can be approximately correlated by the use of conversion tables.

which it relates. However, many millions of tests have shown that in most cases the test-piece does reflect the average properties of the batch of metal being processed and hence indicates its subsequent behaviour in service.

For high-duty uses, such as in aircraft, or for nuclear power, it is important that *all* metal going into service is given a thorough and complete inspection; non-destructive testing then becomes essential. During the pasty fifty years, advances in the development of reliable scientific instruments have made possible the hundred per cent inspection of metal parts without destroying or damaging them. X-ray apparatus is used for 'shadow' examination of castings and wrought metal shapes, providing they are not too dense or thick in section. Thus light alloys for aircraft components are examined for porosity, and variation in grain size.

Cracks are detected by immersion in penetrating fluid of the paraffin type. After withdrawing the component from the liquid the outside surface is dried and time allowed for any liquid which has penetrated the crack to seep out again and be revealed on the surface. Improvements in this procedure have been made by putting fluorescent chemicals in the fluid and examining the part under ultra-violet light, or covering the surface of the casting with a lime wash or chalk to reveal the seeping fluid.

Where magnetic materials, such as iron and steel, are involved, a range of non-destructive tests based on their magnetic properties are used. One involves putting the part in a strong magnetic field and dusting it with fine particles of magnetic iron oxide. The iron oxide congregates where any intensive changes in magnetic intensity occur, such as in the region of cracks and, to a lesser extent, where surface porosity exists.

With non-magnetic metals, such as copper and aluminium alloys, eddy-current tests may be used to detect surface flaws. This procedure involves the generation of small localized electrical high-frequency currents just beneath the surface of the tube or bar of metal under test. Search coils of copper wire pick up the residual eddy currents and compare them with a standard

uniform specimen. This technique discovers whether metal
tubes are free from internal defects; it is particularly difficult for
the human eye to detect such faults.

The old workshop test of striking a metal object with a
hammer and listening to the sound has been developed into
scientific examination by using ultrasonic waves. These sound
waves are basically in the same frequency range as those used
for submarine detection but they are generated in contact
with the metal surface under examination and are then transmitted through the body of the metal, reflected on the far side,
and picked up again by a search crystal close to the generating
crystal. The search crystal transmits a signal to a cathode ray
oscillograph which reveals 'echoes' if cracks, porosity, or other
cavities are present within the body of the metal. This technique
is used in the aircraft industry for the examination of metal parts
before they go into service, for example spar booms in aircraft
wings. Die blocks for plastic injection moulding and diecasting
are examined in this way to determine whether the steel is free
from internal flaws, thus preventing the waste of man-hours that
would occur if a defect in the die block were not revealed till
the last machining operation.

FATIGUE FAILURE

So far, the tests which have been described indicate how sample
pieces of metal behave when they are stressed *once only*, whereas
in service metals often have to undergo thousands, sometimes
many millions, of reversals of stress. For example, the shaft of a
gas turbine aero engine may be revolving at 14,000 to 16,000
times per minute, and a journey of five hours' duration would
mean that the stresses in the metal alternate or fluctuate over
four million times. The turbine blades themselves also vibrate
in a complex manner; frequencies of up to a million per minute
have been recorded, although the stresses involved were
small.

The ordinary tensile test is not necessarily a criterion of the
capacity of a metal to stand up to such repetitions of alternating
stresses. This was realized over a hundred years ago, when

bridges of wrought iron and, later, steel were replacing stone and brickwork bridges. It became essential to learn about the capacity of metals to undergo many repetitions of stress, and in the early 1860s, at the request of the Board of Trade, Sir William Fairbairn carried out some tests in which a load was raised and lowered on to a large wrought-iron girder. It was calculated that the application of a single load of 12 tons would be required to break the girder, but Fairbairn found that if a load of a little more than 3 tons were applied 3,000,000 times the girder would break. He concluded, however, that there existed a certain maximum load, under 3 tons, which could be applied an indefinite number of times without fracture occurring. A few years later the German engineer, August Wöhler, carried the work further, and since that time the 'fatigue' of metals has been intensively studied, particularly since the First World War. Among scientists in Britain, Dr Herbert J. Gough and his fellow-workers have played a prominent part in this branch of metallurgy, which is so important in our era of high-powered engines and fast-moving aircraft.

The method developed by Wöhler for testing the fatigue of metals is extensively used today. A specially shaped test piece is gripped at one end, while at the other end of the specimen a ball race is fitted, from which a load is suspended. The test piece is then very rapidly rotated from the gripped end by a fast revolving electric motor so that under the action of the overhung load it is alternately stressed in tension and compression once each revolution. A mechanism for counting the number of revolutions or reversals of stress, often amounting to 50,000,000 or more, is attached to the machine, which is kept running until the specimen breaks, perhaps after a short time. The load and the number of reversals of stress are both noted; the same procedure is then carried out with a similar test piece of the same metal, but with a smaller load. This time the number of reversals of stress necessary to cause failure is greater. From a knowledge of the load and the dimensions of the test piece the amount of applied stress is calculated; in the type of test we have been considering the stress is expressed with a plus and minus sign, since there has been a certain amount of tension and an equal amount of compression. From the results of a number of such tests on steel a graph similar to *Fig. 35* is obtained.

It will be noticed that when the stress range is high a comparatively small number of alternations of stress are necessary to break the metal, but as the range of stress becomes lower a rapidly increasing number of reversals can be endured until, when the stress alternates from 385 newtons per sq. mm compression to the same amount of tension (the stress alternates about a mean of zero), the graph has flattened out, indicating that this range of stress could be applied an indefinite number of times.

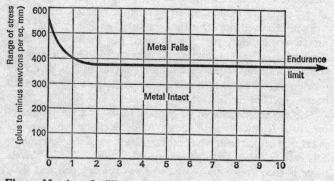

Fig. 35. Number of millions of reversals

Such a steel is said to have an 'Endurance Limit' of plus to minus 385 newtons per sq. mm. If a slightly higher stress of, say, plus to minus 400 newtons per sq. mm were applied, the life of that steel specimen would be only 2 million reversals; parts of a motor-car may be subject to more than that number of alternations of stress within a year of normal usage. In the case of a bridge, however, heavy stresses may be applied only a few times daily, for example, every time an express train passes over; so from this point of view such a bridge could be designed on the basis of a few hundred thousand applications of stress, knowing that this would represent more than the expected life of the bridge.

The above is the 'classical' conception of fatigue limit in steels; but at elevated temperatures, critical for each alloy and metal, there is no definite fatigue limit. Fortunately steels have a fatigue limit at room temperature, but many non-ferrous alloys do not show a definite endurance limit, but have a given life for any given stress range at ordinary temperature. *Table 9* shows the number of reversals endured by 'Duralumin' for varying ranges of stress. It is possible to establish what is a safe load for aluminium and other non-ferrous alloys, based on a knowledge of what a part has to do in service, particularly if its anticipated length of service will elapse before the load under which it works would cause failure. Because of this, connecting rods in racing motor-cycles or cars are allowed a working life of only a few hours so that the number of reversals at the high stress range experienced will be less than the Endurance Limit. This 'finite' life of light alloy components partly accounts for the regular overhauls that are made to aircraft after a given number of flying hours, when some parts are renewed, partly on account of wear, distortion, or corrosion, but also so that they will not be stressed a sufficient number of times to make them liable to fatigue failure.

Table 9. The Endurance of a Duralumin type of Alloy under Fatigue at varying Ranges of Stress

Stress range *Newtons per sq. mm*	*Number of reversals endured before breakage*
Plus to minus 190	1,000,000
Plus to minus 160	5,000,000
Plus to minus 150	10,000,000
Plus to minus 123	50,000,000
Plus to minus 118	100,000,000

There is a 'popular fallacy' about fatigue which should be mentioned here. When a part fails in this way some people remark confidently 'the metal has crystallized'. This comment is based on the fact that the fatigue fracture of metal is different from that shown in *Fig. 34*; in fact the metal seems to have broken off short without any sign of ductility. This is, however, a peculiarity connected with the mode of fatigue fracture; the metal has suffered no inherent change.

Sometimes metal parts which are subjected to fatigue stresses in service fail prematurely under repetitions of stress lower than those established as safe by experiment. One of the likely causes of this is the influence of an abrupt change in section. One has only to recollect the tragic failure around the window frames of the cabin of the early Comet jet aircraft to appreciate the difficulty of ensuring the complete avoidance of design details which may contribute to low fatigue strength in a structure. In a similar manner, scratches, dents, or even inspectors' stamp marks may lead to early failure under alternating stresses. The effect of surface finish is so important that connecting-rods of high-powered aircraft engines are given a super-polished finish in order to obtain the maximum fatigue strength of the part. Another aspect of fatigue failure concerns the influence of corroding media such as salt water exerted at the same time as the fatigue stress; this combination of fatigue and corrosion may lower the endurance limit of metals. Even the exclusion of atmospheric oxygen can raise the fatigue endurance limit of steel.

CREEP

When metals such as steel are used at high temperatures under uninterrupted stresses, as, for example, furnace and steam-boiler parts, they yield very slowly, so that over a period of months or years, they stretch and may eventually fracture; this effect is generally referred to as 'creep'.

Because it is generally at high temperatures that creep is most troublesome to the engineer, this has led to the wrong notion that it is purely a high-temperature phenomenon. However, creep can take place even at low temperatures, particularly in soft metals of relatively low melting point. Thus lead sheets for church roofing thicken near the eaves.

When stress is applied to a metal under creep conditions, the following stages occur:

(1) An initial instantaneous small strain, called micro-creep.

(2) A period during which strain or flow occurs, at a decelerating rate; this is known as the primary stage of creep.

(3) A prolonged period during which further deformation is small and steady. This is known as the secondary stage of creep.

(4) The creep rate accelerates, the test piece elongates rapidly and ultimately ends in fracture. This is known as the tertiary stage of creep.

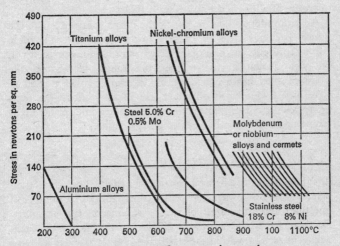

Fig. 36. Stress versus temperature for rupture in 1,000 hours

The best way of determining the suitability of a metal for service at elevated temperature is by means of creep tests. A specimen similar to a tensile test piece is subjected to a constant load, surrounded by a furnace to maintain the entire test piece at the temperature. Accurate measurements are made of the increase of length, over periods of time which may be as long as one to ten years.

In many cases today metals and alloys are not called upon to endure prolonged exposure to high temperatures for years on end. Many aircraft engine components endure high temperatures for only 1,000 to 5,000 hours. This aspect has led to the development of the stress rupture test, to indicate the stress which will cause an extension at one millionth of a centimetre per centimetre of test piece per hour when exposed to a constant temperature for a period of, say, 100, 300, 600, and 1,000 hours. By plotting these stresses a type of curve is developed as shown in *Fig. 36*. This indicates the breaking stress, for different temperatures, causing rupture after 1,000 hours. The curves to the right of the diagram are in block form, to illustrate the behaviour of groups of alloys of the same family types.

An aluminium alloy, for which the curve is at the left of the figure, becomes weak at quite a low temperature. The next curve represents the best obtainable from titanium alloys; it will be noticed that in the temperature range 540°–650°C the titanium alloy has a creep rupture strength similar to that of steel, yet its density is only half that of steel. At higher temperatures the curves show stainless steel and some special high-temperature alloys such as are used in jet engines.

THE MECHANISM OF FAILURE

A study of the mechanical testing of metals shows that their behaviour is not always apparently consistent, and a number of questions come to mind concerning the mechanism of failure. For example, what leads to the difference in behaviour of metals when stressed (a) within the elastic limit, and (b) above it? Does the fracture of metals occur by separation along the grain boundaries or are the grains torn in half? In dealing with these questions much help has been given by studying metals under the microscope during and after stressing and by X-ray examination of the minute distortion of their atomic structure.

If a polished and etched metal such as aluminium or copper is examined under the microscope while being stressed at room temperature, a widespread change appears when the elastic limit is passed. The surface of the metal can be seen to have roughened and each grain is marked with a number of fine parallel lines, the directions of which vary from grain to grain (*Fig. 37b*). In some of the grains two or three series of these lines may have developed. They are called 'slip bands' and are actually steps produced on the surface; they are an outward sign of the permanent deformation of the metal grains.

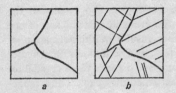

a   *b*

Fig. 37

The mechanism of the formation of the slip-bands can be appreciated from *Fig. 38*. Imagine part of a grain of metal to be stressed as indicated by the arrows in the upper drawings. At first the grain can distort elastically and if the stress is removed it will revert to its original form as shown in the lower drawings on

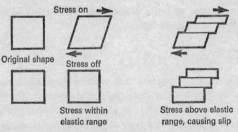

Stress on

Original shape   Stress off

Stress within elastic range

Stress above elastic range, causing slip

Fig. 38. The mechanism of slip

the left. But above a certain stress a part of the grain slips like a pile of coins being pushed over slightly, and when this occurs part of the deformation is permanent, as shown in the bottom right-hand drawing. The planes at which the slip occurs could be observed under the microscope in the form of slip-bands, as shown in *Fig. 37b* and Plate 24c.

As the metal is stressed, the number and the intensity of slip-bands increase until, with sufficient distortion, the structure is so confused that the original boundaries of the grains almost lose their identity. When this state of affairs is reached the metal commences to distort in another manner, by general plastic movement.

In the early stages of deformation the slip results in the hardening of the metal. It appears probable that the slip of one block of the grain over another results in the formation of minute fragments or 'crystallites', and it is doubtful whether slip occurs again exactly in that place because these crystallites have locally strengthened and hardened the metal. The arrangement of the atoms in a grain of metal is such that certain planes exist along which slip can take place more easily than in other directions.

The research that has been done on dislocations, previously discussed on page 85, has thrown a great deal of light on the reasons why metals are so strong, and how and why they fail under increased stress. It is probable that the inspiration which led to metals being strengthened by minute fibres of other materials was derived from the general spread of knowledge of the effect of dislocations, and the realization that such added fibres can produce increased strength.

THE BREAKAGE OF HOT METALS

It will be remembered from page 72 that when a stressed metal is heated above a certain temperature the metal re-crystallizes. When the metal is stressed while it is above this temperature slip-bands are not observed, but a type of plastic flow (possibly different from that at room temperature) occurs from the start and often results in a considerable deformation of the metal, which usually fails by cracking around the grain boundaries. The major difference between the rupture of metals when hot and when cold is that in the former state the grain boundaries are the weakest parts, while in the latter the grains themselves are the weakest.

CHOOSING THE RIGHT METAL FOR THE RIGHT JOB

Table 10 shows some of the mechanical properties of some well-known metals and alloys. The strength has been expressed in the old tons per square inch and in SI units. In selecting a suitable alloy for a given purpose, designers and metallurgists should work in close co-operation, bearing in mind that a wide choice of metals and alloys of varying strengths are available. Lead has a tensile strength of only 15 newtons per sq. mm while, as is seen from *Table 10*, heat-treated alloy steels are nearly a hundred times stronger.

The strength of a part increases in proportion to its effective section, but this way of increasing strength is not always desirable, as it increases the dead weight. Even in a bridge the greater part of the strength is used in supporting its own weight; increase of section augments the load which the bridge has to support. The correct procedure in designing any component is to arrange that the metal shall be thick at just those places where strength is required; this is why girders used in building construction are flanged, for more metal is disposed at the top and bottom, where the stresses are greatest.

A metal which is ideal for enduring one kind of stress may be unsuitable for another; cast iron, for example, is strong in compression, but weak in tension. Thus it is essential to determine the type and size of the stresses to be met and then select the material most able to meet that stress or combination of stresses. A railway coupling needs to be made in a strong but ductile metal (in other words, one with a high elongation figure), so that any sudden overload can be absorbed without fracture occurring; a brittle metal would not be suitable, however strong, since the smallest stretching due to severe overload, such as might sometimes arise, would cause immediate breakage.

Not only must the metal be fit to withstand the stresses applied in service, but it must be capable of being shaped into its finished form without difficulty. So the task of the manufacturer must be envisaged, including casting, forging, pressing, machining, and heat-treatment. This aspect is unfortunately

Table 10. Some Mechanical Properties of well-known Metals and Alloys
(These figures are approximate)

Metal or Alloy	Condition	As used for	Brinell Hardness Number	Tensile strength		0·1 per cent proof stress		Elongation per cent on 56 mm
				tons per sq. inch	newtons per sq. mm	tons per sq. inch	newtons per sq. mm	
Aluminium	Wrought and annealed	Frying pans	27	6	92	2	31	18
Aluminium alloyed with 7% magnesium	Wrought and annealed	Tubes and sheet for aircraft	80	20	309	8	124	17
Duralumin	Wrought and heat treated	Aircraft	115	28	434	17	264	15
Magnesium alloy, with 8% aluminium	Cast and heat treated	Aircraft landing wheels	60	17	264	5	77	10
Copper	Wrought and annealed	Tubes	50	14	216	4	62	55
Copper	Cold-drawn into wire	Copper wire	110	28	434	26	403	4
70/30 brass	Deep-drawn	Cartridges	160	35	540	30	463	10
70/30 copper–nickel alloy	Drawn into tube	Condenser tubes	170	38	587	30	463	8
Mild steel	Hot-rolled into plate	Ships' plates	130	30	463	15	232	25
Alloy steel with 3·7% nickel 0·8% chromium 0·2% carbon	Forged, quenched and tempered at 400°C	Camshafts	400	86	1,340	77	1,188	14
	Forged, quenched and tempered at 600°C	Gears	300	65	1,004	57	880	22
Cast iron	Cast	Lathe beds	200	14*	216	—	—	—

* It would seem that when lawyers speak of 'a cast-iron case', they are displaying little appreciation of the strengths of metals. 'An alloy steel case' would be more appropriate.

not always sufficiently appreciated. Other special considerations of duty may have to be considered. The metal may have to work in corrosive conditions, or may require some specific magnetic or electrical properties.

Cost is an important item. Platinum might be a very suitable metal for a number of applications, but its high cost limits its use.

10

IRON
AND STEEL

A recent discovery illustrates that nineteen hundred years ago iron nails were made and used on a considerable scale. Inchtuthil, a Roman site near Perth in Scotland, was built in A.D. 83 as the advance headquarters of Agricola. The legion was withdrawn after occupying the site for only six years and was told to leave nothing that could help the enemy. Timber was removed, pottery smashed, and wattle burned; but one valuable load, weighing about seven tons, could not be removed; it consisted of 763,840 small, 85,128 medium, 25,088 large and 1,344 extra large nails. The small ones were 50 mm long and the extra large were magnificent tapered spikes 400 mm long, square in section, with solid heads. The Romans dug a deep pit in the corner of the store, poured the nails into it, packed two metres of clean earth on top, demolished the building, and removed all traces of the cache.

In 1961 the late Sir Ian Richmond of Oxford discovered and unearthed the nails. Most of them were completely corroded, but those in the centre of the mass (about one per cent of the whole) were remarkably well preserved, retaining their clean-cut heads and edges, even after burial for nineteen hundred years. This unusual find was carefully studied by experts in early metallurgy, who reported that the metal was heterogeneous in composition, varying from almost pure iron to high-carbon steel. This supported the view that early smelting processes produced a spongy mass of iron which had never been molten during the whole operation. This sponge was repeatedly heated and hammered to consolidate it and to expel entrapped slag. For the manufacture of nails, bars were forged, which were then cut and tapered at one end. The shaft was then held in a die while the head was forged.

The Latin word for iron was *ferrum*; iron, steel and cast iron are classed as 'ferrous metals', indicating that they consist largely of iron and distinguishing them from the other, 'non-ferrous', metals. The extraordinary variety of properties and uses of the ferrous metals is made possible by the effect of varying amounts of carbon and some other elements when alloyed with iron.

The ferrous metals are at the root of our material civilization. Without them there might have been no great liners, no sky-scrapers, railways, motor-cars, tanks, or tractors. If this book had been written with the space devoted to each metal pro-portionate to its tonnage, more than 300 pages would have been about iron and steel and only 20 pages devoted to all the other metals. As illustration of the scale of manufacture of iron and steel, ten large blast furnaces can produce in one year over fifteen million tons of pig iron; this is more than the whole world's annual output of aluminium. The Forth Bridge and the Sydney Harbour Bridge each contain over fifty thousand tons of steel. In Britain alone over ten million tons of iron and steel are used in our railway-line systems and two million tons in loco-motives.

The following table shows the world production from 1952 to 1975, and highlights the tonnage made in Britain and in the three countries which are now the major producers of steel. The trebling of the world's production is noteworthy enough, but the emergence of Japan has been extraordinary.

Table 11. Steel Production in Millions of Tonnes

Country	1952	1962	1972	1973	1975
U.S.A.	84·52	89·20	120·88	136·46	108·41
U.S.S.R.	34·50	76·31	126·03	131·00	140·20
Japan	6·99	27·55	96·90	119·32	102·21
Britain	16·68	20·82	25·32	26·65	20·01
(World total)	211·6	358·3	628·49	694·31	642·70

Steels may contain up to about 1·5 per cent carbon, though the grade of steel generally used contains only about 0·2 to 0·3 per cent. Several elements beside carbon are present in steel; some, like manganese, have a beneficial effect; others, for example sulphur, may be harmful; steel-makers reduce the amount of such harmful impurities as much as is economically possible. The addition of nickel, chromium, molybdenum or tungsten produces 'alloy steels', including high-speed steels, stainless steels, and die steels. Cast iron contains between 5 and 10 per cent of other elements including carbon, silicon and manganese. It is produced by remelting pig iron, which is the product of the blast furnace and is the cheapest of all metals. Alloy cast irons contain alloying elements such as nickel or chromium.

The steels are by far the most important ferrous metals, followed by cast iron and finally wrought iron and other low-carbon irons. A hundred years ago the position was quite different; steel was an expensive material, produced in only small quantities for such articles as swords and springs, while structural components were made of cast iron or wrought iron. Thus, the first metal bridge in Europe, at Ironbridge over the River Severn, was erected in 1799 and embodied nearly 400 tons of cast iron. The Eiffel Tower was made of about 7,300 tons of wrought-iron girders, by Forges de Wendel.

Wrought iron, which was used for other purposes, including chains and entrance gates, was made by a laborious process of re-fining pig iron. Although the temperature of the furnace was sufficient to melt pig iron (melting point about 1,200°C) it was not high enough to keep the pure iron molten (melting point about 1,500°) so the refined metal had to be extracted from the furnace in white-hot spongy lumps, after which it was 'wrought' – forged and rolled to the required section.

THE BESSEMER PROCESS

In August 1856, an Englishman, Henry Bessemer, made public the description of a process which eventually reduced the price of steel to about a fifth of its former cost and, more important still, made it possible to produce steel in large quantities. This was a mainstay of the steel industry for over a hundred years and when, in the early nineteen sixties, it began to be replaced rapidly by the oxygen process, the principles involved could not have been established without the experience gained during the previous hundred years.

Bessemer proposed burning away the impurities by blowing air through molten pig iron, an idea that appeared fantastic and dangerous to the Victorian iron-makers. However they invested large sums of money in his process, only to be scandalized that they could not make it work. Bessemer paid back all their money, spent several thousand pounds in discovering what had gone wrong, and proved that his own experiments had been with an iron containing only a small percentage of phosphorus; they had unfortunately tested his converter with cast irons of a high phosphorus content. He then tried to persuade his disillusioned clients to try his process again but they had been 'once bitten'. Bessemer then decided to go into production himself, built his own steel works in Sheffield and was soon making nearly a million tons of steel per annum. He continued to meet opposition and once, when he attempted to interest a railway engineer in the possibilities of steel for railway lines, received the reply 'Mr Bessemer, do you wish to see me tried for manslaughter?'.

The Bessemer 'converter', shaped like a huge concrete-mixer, is mounted in such a way that it can be tilted to receive a charge of twenty-five to fifty tons of molten pig iron; it is then brought upright for the 'blow' to take place (*Fig. 39*). Air is blown through a number of holes in the base of the converter and forces its way through the molten metal. The oxygen from the air blast combines with some of the iron, producing iron oxide which dissolves into the molten metal and then reacts with the silicon, manganese, and carbon, which thus become oxidized. The oxides of iron, manganese, and silicon combine to form a slag, while the carbon is removed as carbon monoxide, part of which burns and forms carbon dioxide. The time taken for the removal of impurities is about 15 minutes and the complete operation, from one tapping to the next, occupies 25 to 30 minutes. No external fuel is applied and the whole of the heat requirement is furnished by the oxidation of the impurities, together with that of some of the metallic iron. The converter is tilted and its metallic contents are poured into a large ladle.

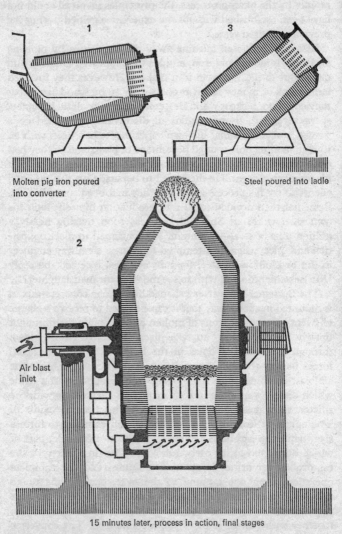

1

Molten pig iron poured into converter

3

Steel poured into ladle

2

Air blast inlet

15 minutes later, process in action, final stages

Fig. 39. The Bessemer converter

An alloy containing manganese is added while the metal is being poured into the ladle; the manganese combines with dissolved iron oxide, thus removing it from the steel. Other additions such as ferro-silicon and aluminium are made to assist the deoxidizing of the steel, then carbon in the form of anthracite is added to bring it to the correct carbon content.

The process described above was known as the 'acid' Bessemer process: the converter was lined with silica bricks and because these are known in the refractory trade as 'acid', to differentiate them from 'basic' refractories containing oxides of metals, the process got the name, which of course does not imply that acid was used. The acid Bessemer process, as the Victorian ironmasters discovered, could not eliminate phosphorus, which is harmful to steel, so low-phosphoric pig irons had to be used. In 1878, two Englishmen, Sidney Thomas and Percy Gilchrist, contributed an improvement whereby they lined the converter with 'basic' refractory bricks, containing magnesia or dolomite. Lime was added to the bath to combine with the phosphorus and silicon, and thus remove them from the iron in the form of slag containing calcium phosphate and calcium silicate. This caused the weight of slag and the amount of heat required to be higher than in the acid Bessemer process. To provide this heat the amount of phosphorus in the iron needed to be between 1·8 and 2·0 per cent. This impurity, which was being chemically transferred from the iron to the slag, therefore helped to provide enough heat of reaction to make the process possible. The basic lining of the converter was to provide conditions under which the reactions with the lime could take place without destroying the furnace lining. If silica brick were used, as in the acid process, the lime would attack it chemically.

Since air, a mixture of nitrogen and oxygen, is used, the resulting steel contains nitrogen which makes steels liable to brittleness. Furthermore the bulk of the nitrogen which is not dissolved carries away so much heat that only a metal of high phosphorus content will generate enough heat to give the required temperature of the liquid steel in the basic Bessemer process. The remedy was to replace the air blast by oxygen or a gas mixture containing no nitrogen; this overcame both the difficulties at the same time, while producing the steel more economically. Such developments led to the recent revolutions in steel-making processes, described later.

Twenty-three years after the introduction of the Bessemer process a tragedy occurred which caused attention to be drawn to the possibilities of construction with steel. On the night of 29 December 1879, a bridge over the River Tay, in Scotland, collapsed while a train was crossing, and seventy-eight lives were lost. The bridge, built of wrought iron and cast iron, with 84 spans each 70 metres in length, had been counted one of the engineering wonders of the world. The failure of the structure was due not so much to low strength in the wrought iron as to faulty manufacture of cast-iron columns and inadequate allowance for the stresses of wind and flood which were so great on that stormy December night. Nevertheless, the disaster stimulated engineers to reconsider the question of the materials of construction; it was realized that steel, which by that time was being produced on a considerable scale and fabricated into standard shapes, was a more suitable engineering material than wrought iron.

THE OPEN-HEARTH PROCESS

A naturalized Englishman, of German origin, developed the process which became widely used for steel making, and which still accounts for a very considerable tonnage. In the late 1850s, Dr Charles W. Siemens was concerned about the relatively small quantities of heat obtained in the existing furnaces using solid fuels such as coal and coke. He and his brother Frederick and, later, the Martin brothers of France, experimented with gaseous fuels and developed a method of 'heat regeneration', making the outgoing hot burnt furnace gases pre-heat the incoming gaseous fuel. By this means a sufficiently high temperature was obtained to treat large quantities of metal and to keep it molten throughout the process, enabling it to be cast into ingots when the refining was complete.

The Siemens–Martin 'open-hearth' furnace is so called because the molten metal lies in a comparatively shallow pool on the furnace bottom or hearth. *Fig. 40* shows the furnace, with the heat-regenerating chambers at a lower level. These contain firebrick 'chequerwork', arranged so that the hot exhaust gases from the furnace pass through the channels so formed, and transfer heat to the firebricks in one set of regenerators. In the meantime the gaseous fuel and air are being passed separately through the other regenerators, which have previously been heated. The directions are reversed at intervals, to maintain a high temperature of combustion. Pig iron and steel scrap are charged into the furnace from the platform on the right; when the steel has been made it is tapped into the ladle shown on the left.

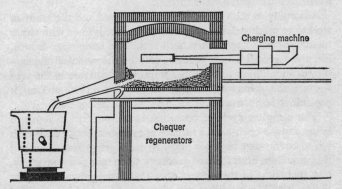

Fig. 40

Open-hearth units holding 100 to 300 tons of metal are common and some of over 500 tons' capacity are in use. Often the furnace is designed to be tilted for slag removal and for tapping the molten steel. If the blast furnace plant is adjacent to the steel-works, the molten pig iron is transferred in ladles, holding 40 to 100 tons of liquid metal, to 'mixers'; these are large barrel-shaped tilting furnaces, often of more than 1,000 tons capacity, in which the pig iron can be stored temporarily and kept molten, ready for conversion into steel.

Usually scrap steel is previously charged and heated in the furnace and the liquid pig iron added to this; thus the impurities in the pig iron are 'diluted' and the refining process does not take so long as if the entire charge were pig iron. Very often more than half of the charge consists of scrap. Then iron oxide, in the form of iron ore or scale, is added; this, together with oxygen in the furnace gases, oxidizes the impurities, the carbon being removed as carbon monoxide. The silicon and manganese are also changed into their oxides, and these react with added sand or lime to form a slag which is removed separately. At the end of the process, when the impurities have been brought down to the required level and the metal has been tapped, additions of ferro-manganese and ferro-silicon are made to bring the steel to the correct composition; later a small addition of aluminium is made to deoxidize the metal further.

According to the type of pig iron employed and the grade of steel to be produced, the open-hearth furnace is lined with either 'acid' or 'basic' refractory firebricks. The basic open-hearth, like the basic Bessemer process, permits the removal of phosphorus in addition to the other impurities, whereas in the acid furnace the refractory linings are of silica brick, making it possible to remove only carbon, silicon, and manganese.

The complete cycle of operation lasts from five to fourteen hours and the open-hearth furnace of average capacity deals with ten to twenty-five tons per hour. Remembering from chapter 3 that a modern blast furnace produces 5,000 tons of iron per day, and that some of the newest ones make more than twice that figure, it will be seen that one blast furnace could keep at least eight open-hearth furnaces in operation. This rather inefficient performance of the open-hearth compared with the enormous output of the blast furnace justified the emergence of the highly productive oxygen process for steel. The operating cost of the open-hearth furnace is greater than that of the Bessemer converter and substantially more than that of an oxygen furnace.

THE USE OF OXYGEN IN STEEL-MAKING

The refining of steel by Bessemer and open-hearth processes removes impurities from pig iron by iron oxide and by the oxygen of the air, most of the impurities being taken into the slag. It had long been realized that the replacement of air by high-purity oxygen offered the prospects of striking improvements in steel-making, but the high cost of oxygen was an obstacle to this development. In the 1960s steel-making took a great leap forward, thanks to the production of oxygen on such a scale that it is measured by the tonne (about 700 cubic metres) and at a fraction of its former cost. Plants have been built adjacent to the larger steel-works, each capable of providing several hundred tonnes of high-purity oxygen a day.

The first developments of the oxygen process began in Austria, but soon other Continental steel-makers followed suit. They had depended to a large extent on Bessemer production of steel and it was possible for them to gain confidence in the use of oxygen by at first enriching the air blast with some oxygen, or by a mixture of steam and oxygen, or carbon dioxide and oxygen. Within a few years several processes were invented, three of which are described below.

The L-D Process

The title of this process is an abbreviation of 'Linz Düsenverfahren', or Linz lance process. It is however often suggested that the name was derived from the initials of two separate plants in Austria, at Linz and Donawitz. The local Austrian ore was too low in phosphorus to enable the air-blown basic Bessemer method to be used, while the amount of steel scrap available was not high enough to make open-hearth steel economic. Thus a combination of circumstances in Austria provided the incentive for this major development, which radically altered the steel industry all over the world.

The L-D process consists of blowing a jet of almost pure oxygen at high pressure and travelling at supersonic speed on to the surface of molten iron, held in a converter as illustrated in *Fig. 41* and Plate 13. The vessel remains stationary in a vertical position throughout the blow. This process combines the low capital cost and speed of operation of the Bessemer process with the high quality of the open-hearth. Furthermore about 20 per cent scrap can be added to the charge. The high speed of the reaction makes chemical analytical control difficult and the

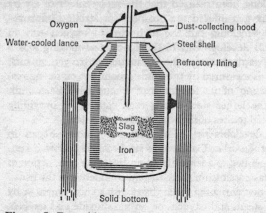

Fig. 41. L–D top-blown converter

installation of semi-automatic 'press button' analysis machines is an important aid. The original L–D process was limited to iron whose phosphorus content was under 0·3 per cent. Some years ago, it was discovered by French and Luxemburg steel-makers that injection of lime with the oxygen permitted the use of iron of much higher phosphorus content – up to nearly 2 per cent. This version of the process is known as the LD–AC or BOF (Basic Oxygen Furnace), and has enabled steel producers to treat a much wider range of irons, produced from imported and indigenous ores.

Price differentials between liquid iron and cold scrap have motivated steel-makers to provide extra heat to melt more scrap than the process could handle under normal operating conditions. Among these methods are the application of an oil fuel to the oxygen, the addition of calcium carbide, and silicon carbide.

Other Oxygen Processes

The Kaldo process was pioneered by Professor Bo Kalling at Domnarfvet, in Sweden, from which the name Kaldo was derived. The converter consisted of a vessel of similar pear-shape to that of the L–D converter; it rotated on an inclined axis during the blow, at speeds of up to thirty revolutions per minute. As the capacity of the converter was 125 tons, the mechanical problems of rotating the vessel at such a speed can be imagined. An oxygen jet was blown against the surface of the bath; variations in the rate of oxygen supply and the speed of rotation, together with the relatively longer time occupied (about ninety minutes for the full cycle from one tapping to the next), permitted a closer control of composition than in the L–D process. However, after a number of years it was found that the high maintenance and refractory costs involved in the Kaldo process made it uneconomic; its use has been practically discontinued.

There is an important distinction between the L–D and Kaldo processes. L–D uses a supersonic jet of oxygen to punch the gas into the metal, to emulsify and foam the slag where the major part of the steel-making takes place. The objective of the Kaldo process was to flood the furnace atmosphere with oxygen and burn carbon monoxide from the steel-making reaction within the vessel, thereby enhancing the capacity of the process to melt solid scrap.

The Rotor process was developed by Dr Rudolf Graef at Oberhausen. The converter, illustrated in *Fig. 42*, is capable of turning at one revolution in two minutes. A primary jet blows oxygen into the bath of metal and a secondary jet above the bath converts carbon monoxide into carbon dioxide, thus ensuring heat economy. It is possible to produce high-grade steels from

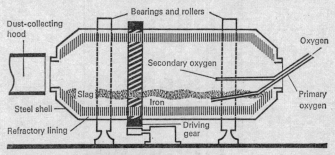

Fig. 42. Rotor mixed-blown converter

Table 12. Steel Tonnages per Annum produced in Britain

Year	Acid Bessemer	Basic Bessemer	Acid open hearth	Basic open hearth	Oxygen	Electric	Others	Total
1870	215,000	—	—	—	—	—	—	215,000
1880	1,034,000	10,000	251,000	—	—	—	—	1,295,000
1890	1,613,000	402,000	1,463,000	101,000	—	—	—	3,579,000
1900	1,254,000	491,000	2,863,000	293,000	—	—	—	4,901,000
1910	1,138,000	641,000	3,016,000	1,579,000	—	—	—	6,374,000
1920	587,000	375,000	3,380,000	4,580,000	—	89,000	56,000	9,067,000
1930	279,000	—	1,805,000	5,099,000	—	76,000	66,000	7,325,000
1940	176,000	738,000	2,174,000	9,274,000	—	435,000	178,000	12,975,000
1950	248,000	845,000	1,311,000	12,981,000	—	736,000	178,000	16,293,000
1960	294,000	1,655,000	658,000	19,875,000	25,000	1,686,000	112,000	24,305,000
1970	283,000	—	162,000	13,213,000	9,102,000	5,517,000	39,000	28,316,000
1972	223,000	—	61,000	9,300,000	10,786,000	4,914,000	36,000	25,320,000
1974	139,000	—	20,000	6,168,000	10,797,000	5,271,000	30,000	22,425,000

pig irons with up to 2 per cent phosphorus. The original Rotor converter was of 60 tons' capacity but later units at Oberhausen, Peine, and two plants in South Africa are of 100 tons' capacity, each capable of producing 500,000 tons of steel per annum.

Table 12 shows the rise and fall of the Bessemer and open-hearth process during the last hundred years and the emergence of oxygen processes since 1960 in Britain. Japan is the leading exponent of oxygen steel-making, with an annual capacity of over 80 million tonnes. America has made substantial increases in her oxygen steel capacity. For example, in 1970 the U.S., with a total steel production of about 131 million tonnes, produced 48 per cent by oxygen, 37 per cent by open hearth, and 15 per cent by electric process. By 1972, with approximately the same total production, the percentages had changed to 56, 26, and 18 respectively. This was accompanied by a 10 per cent increase in productivity during those two years.

ELECTRIC-FURNACE STEEL

For high-grade alloy steel cutting tools, die steels, and stainless steel, the metal must be refined and melted under rigidly controlled conditions and in such a way that impurities are reduced to a minimum. Where a fuel is burnt in the furnace some contamination is unavoidable; and this led steel-makers to realize that electric melting was likely to be technically more desirable than the methods of the open-hearth and Bessemer processes. The electric furnace was originally intended chiefly to refine and produce alloy steels of good quality. Pig iron is not treated directly in electric furnaces, though sometimes it is partly refined in an open-hearth furnace and then transferred to electric furnaces for final treatment and alloying.

The electric-arc furnace is illustrated in *Fig. 43*. The hearth can be either acid or basic lined. Acid furnaces are mainly used in steel foundries and are rarely of more than 10 tons' capacity. Basic furnaces of up to 80 tons' capacity are now used for making alloy and special steels. The bottom of the furnace is covered with lime; scrap steel of known quality is then put inside. Next the three carbon electrodes are automatically lowered to the

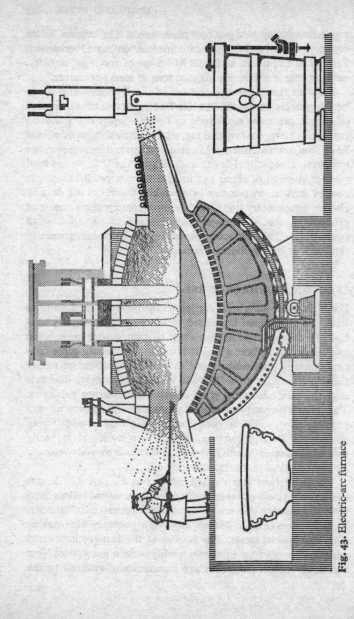

Fig. 43. Electric-arc furnace

surface of the metal and melting begins. When melting is complete the slag will already have removed much of the silicon, manganese, and phosphorus from the molten scrap. Iron ore is added for the removal of carbon and the remainder of the phosphorus.

Next the furnace is tilted and the slag raked off into the ladle on the left of the diagram; it is replaced by a slag, composed of lime, fluorspar and carbon, which removes sulphur from steel. A sample of the steel is analysed and adjustments to the composition are made by adding ferro-alloys. Finally the temperature is checked, the furnace tilted, and the metal tapped into the ladle on the right of the drawing.

One great advantage of the electric process is its ability to operate entirely on scrap charges and still to produce good quality steel. In the past it has often been stated – erroneously – that the L–D oxygen process is making the open hearth redundant. This is true to only a limited extent because the oxygen process requires a liquid iron charge and can therefore only replace those open-hearth furnaces which worked on a high proportion of liquid metal. A large majority of the remaining open-hearth shops use steel scrap in situations where there is no blast furnace production to provide the liquid metal. These open hearths are being supplanted by electric furnaces which, together with the large oxygen steel capacity, will produce all British steel requirements by 1980. Five areas will become the non-alloy steel-making centres – Llanwern and Port Talbot in South Wales, Ravenscraig near Glasgow, Scunthorpe in Lincolnshire and Lackenby in the Teesside district.

CONTINUOUS CASTING OF STEEL

Not many short cuts are possible in the making and processing of steel but one interesting development, in which European steel-makers have played a large part, is the production of sections direct from liquid metal. Continuous casting of steel was developed in Europe in the mid 1950s and has grown rapidly till the present world capacity for continuous casting is over

100,000,000 tons per annum and is still increasing as the process becomes improved and more new steel plants are built.

The continuous casting of steel has been developed from the similar process, described on page 99, for non-ferrous metals. The liquid steel is poured from a ladle into a water-cooled mould. The mould is open ended, but sealed by a dummy bar. When casting commences, the metal solidifies as it comes into contact with the water-cooled walls of the mould; this results in a cup of steel, solid at the base and sides, liquid in the middle, which is cooled progressively by water sprays as it is withdrawn downwards by the dummy bar. While withdrawal is going on, metal is continually being poured into the top of the mould to keep the level in the mould constant. Each mould consists of a pure copper tube about a metre long and of a size corresponding to the shape of the section to be cast. The tube is surrounded by a water jacket which allows it to be cooled continually by a high pressure water supply.

Early attempts to cast steel by this process were beset with a number of difficulties, the most troublesome of which was the sticking of the solidified steel shell to the walls of the mould. The use of a lubricant and the introduction of a reciprocating movement of the mould overcame this problem. It was found possible to increase casting speeds by arranging that the mould moved downward somewhat faster than the speed at which the casting is withdrawn.

The original continuous casting machines for steel involved pouring the metal from a height of over 13 metres from ground level, with the cast section of metal descending vertically and being progressively water cooled. The operating height was then reduced by bending the section after solidification; a further reduction in height was achieved by casting the section in a curved mould and cooling on the curve instead of vertically. *Fig. 44* shows these various stages in the development of continuous casting machines. The machine illustrated in Plate 11 is compact and highly productive, based on the Swiss 'Concast' designed in Europe by a combination of British, Swiss, German and French firms. Plate 12 shows an eight-strand Concast bloom-casting machine made in Japan for the Kawasaki Steel

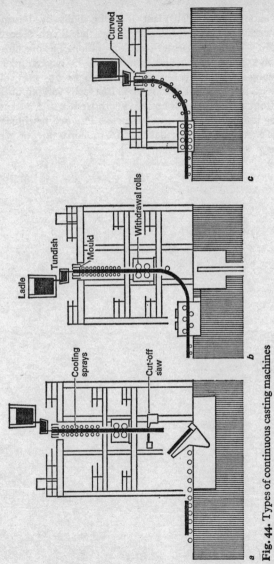

Fig. 44. Types of continuous casting machines

Corporation. A similar plant is being built by Distington Engineering Company, England, for the South Teesside's Lackenby Works of British Steel Corporation. The machine will take in 250/270 ton ladles from the L–D oxygen plant at Lackenby and cast this hot metal into blooms 255 mm square. A second machine at Lackenby will cast slabs varying in thickness from 152 mm to 254 mm, in width from 1 to 2 metres, for cutting into length from 4 to 8 metres. The two machines will have a combined annual capacity of 1·6 million tonnes per year.

11

THE ROLE OF
CARBON IN STEEL

When carbon is alloyed with iron, the hardness and strength of the metal increases; for example, a steel containing 0·4 per cent of carbon may be twice as strong as pure iron, and with about 1·0 per cent carbon, nearly three times as strong, though as the carbon content rises, the ductility is reduced. From about 1·0 to 1·5 per cent carbon the hardness increases further but the strength of the steel diminishes somewhat. Such high-carbon steels are not so much used in engineering as steels with low and medium carbon content, though tools and other instruments which are required to be very hard have carbon contents up to 1·5 per cent. Alloys of iron containing between 1·5 and 2·5 per cent carbon are rarely used. When it contains more than about 2·5 per cent carbon the metal is classed as cast iron, and characterized by good castability, moderate strength and hardness, and usually by low ductility.

For a girder or a ship's plate, a strong but ductile metal is required, capable of being fabricated cheaply and rapidly, and a steel with about 0·2 per cent carbon is employed. For a razor a hard metal is needed which can be brought to an enduring sharp edge, and the razor steel contains over 1 per cent of carbon. For guttering round the eaves of a house a cheap metal is required which can be cast to shape and which has no need to be outstandingly strong, and cast iron is therefore frequently used. So, depending on the application to which the metal is to be put, a suitable iron–carbon alloy is selected which will give the desired properties. *Fig. 45* illustrates some of the uses of steels of various carbon content; cast iron is also shown in this diagram. *Fig. 46* shows some steels which are used in the manufacture of a rifle. Both the carbon (C) and manganese (Mn)

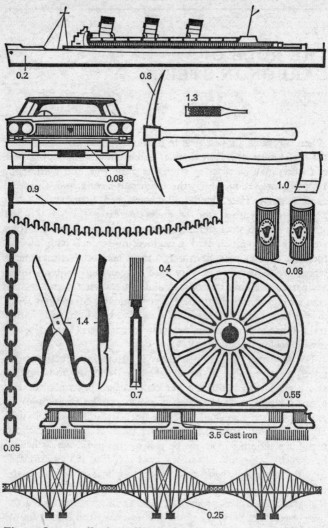

Fig. 45. Some applications of iron and steel
(Figures indicate percentage of carbon)

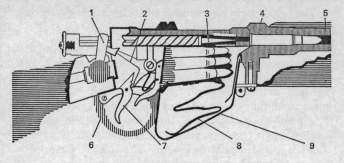

1. Bolt, 0.7% C, 0.4% Mn, H.T.
2. Sear, 0.9% C, 0.3% Mn, H.T.
3. Striker, 0.5% C, 0.5% Mn, H.T. at point
4. Bolt Head, case hardened
5. Barrel, 0.5% C, 0.6% Mn
6. Trigger Guard, 0.35% C, 0.5% Mn
7. Trigger, 0.7% C, 0.9% Mn, H.T.
8. Magazine Spring, 0.75% C, 0.2% Mn, H.T.
9. Magazine Case, 0.18% C, 0.3% Mn

Fig. 46. Some steels in the Short Magazine Lee-Enfield Mark III Rifle

contents are shown, and where a part is heat-treated, this is also indicated (H.T.).

The iron–carbon alloys are usually described according to the amount of carbon present and, though there is no precise demarcation, the following table gives the usual classification:

Table 13

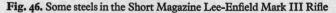

Description	Approximate carbon content, per cent
Mild steel	up to 0·25
Medium-carbon steel	0·25–0·45
High-carbon steel	0·45–1·50
Cast iron	2·50–4·50

The behaviour of carbon steels which have been slowly cooled is attributable to three causes:

(1) At room temperature the atoms of iron are arranged in the body-centred cubic lattice pattern (*Fig. 21*, page 80). When heated to 906°C, the iron atoms spontaneously rearrange into the face-centred cubic lattice pattern (*Fig. 20*). Why this takes place is not fully understood but it is a fact of great importance to a metal-using civilization, for if the change did not occur, steels would not be so amenable to heat-treatment.

(2) At room temperature less than a hundredth of one per cent carbon can exist in solid solution in slowly cooled iron of body-centred atomic arrangement, but at high temperatures up to 1·7 per cent can be taken into solid solution in face-centred cubic iron, this maximum solubility being attained at 1,145°C.

(3) When molten steel containing carbon is allowed to solidify and cool slowly, the carbon does not come out of solution as such, but each carbon atom unites with three iron atoms to form a compound, iron carbide, Fe_3C. Although a compound of a metal and a non-metal, it is similar in behaviour to the general class of 'intermetallic compounds' discussed on page 75 and its presence confers increased hardness.

Carbon dissolved in iron has a pronounced effect on the temperature at which the change in the atomic lattice occurs. Thus if pure iron is gradually heated from room temperature to bright yellow heat, but not melted, the arrangement of the iron atoms changes from body-centred to face-centred cubic at 906°C. With 0·3 per cent carbon the change commences at about 730°C and is complete at about 800°C. With about 0·8 per cent the change begins and completes itself at about 730°C. In fact when carbon is present, the atomic reshuffling on heating commences at 730°C and concludes at that or some higher temperature, depending on the amount of carbon. *Fig. 47* illustrates this, and the figures given above can be checked by reference to the diagram.

The V-shaped graph is reminiscent of that representing the effect of one metal on the melting point of another, where a eutectic (*Fig. 12*, page 64) is formed, which has a minimum

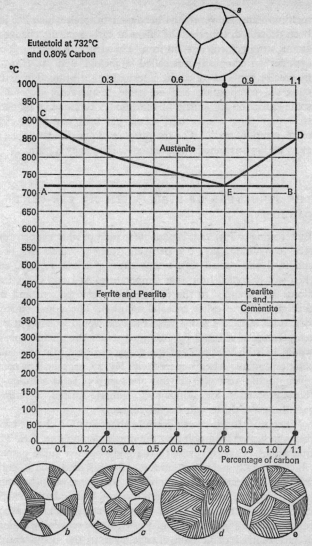

Eutectoid at 732°C
and 0.80% Carbon

°C

Austenite

Ferrite and Pearlite

Pearlite
and
Cementite

Percentage of carbon

Fig. 47

melting point. However, the behaviour of steel, which has just been described, concerns the effect of carbon on the change of atomic arrangement of solid iron. The structure of the 0·8 per cent carbon composition is called 'eutectoid' and 730°C or, to be precise, 732°C is called the eutectoid temperature. The actual eutectoid composition and eutectoid temperature depend to some extent on the purity of the iron. A commercial grade of steel may show a eutectoid at 0·9 per cent and a eutectoid temperature of 700°C. For the purpose of our discussion, however, we have assumed that the iron–carbon alloy is pure.

THE STRUCTURE OF STEEL WITH 0.3 PER CENT CARBON

By referring to *Fig. 47* it is possible to picture what happens to a steel containing, say, 0·3 per cent carbon, which has been solidified and cooled to a temperature of about 1,000°C and then slowly cooled from that temperature. At 1,000°C the iron atoms are arranged in the face-centred cubic pattern and the carbon therefore exists in solid solution. The upper 'microstructure' marked *a* shows what the steel would look like at this temperature if it could be examined under the microscope. It would be seen to consist of the grains characteristic of a solid solution and might be compared with *Fig. 17*, page 73. The name given to the structure of this iron–carbon solid solution is 'austenite' after Sir William C. Roberts-Austen, a famous metallurgist of the late nineteenth century.

When this steel is cooled down to about 800°C a change takes place, indicated by a point on the upper sloping line, CE. The iron atoms begin to revert to the body-centred cubic pattern which normally holds only minute amounts of carbon in solid solution. The carbon atoms do not at once come out of solution but migrate towards areas where the iron atomic lattice is still in the face-centred cubic form, thus increasing the local carbon concentration. Finally, at about 730°C, all the remaining regions of face-centred cubic iron change over to the body-centred cubic arrangement and the carbon can no longer be contained in solid solution. This temperature is shown by the horizontal line, AEB.

By referring to illustration *b*, representing the microstructure of slowly cooled 0·3 per cent carbon steel, it will be seen that the final structure is duplex.

(1) About two-thirds of the structure consists of grains of iron, the white constituent in the illustration. In metallography this is identified as 'ferrite' and it is found that the presence of the grains of ferrite confers ductility on the steel.

(2) About one-third of the structure is a layered formation, the composition of which is discussed below.

PEARLITE

The precipitation of carbon from the solid solution is complex. It is deposited as extremely hard iron carbide, Fe_3C. 'Cementite' is the general term used to identify this constituent, which is deposited in layers arranged alternately with layers of the surplus ferrite; this formation will be seen as the darker constituent in illustration *b*. The layered structure is called 'Pearlite' because, when viewed under the microscope with oblique lighting, it produces an iridescent appearance, like that of mother-of-pearl. This formation was first seen by Professor Sorby when he was developing the use of the metallurgical microscope, in the early 1860s. A photograph of pearlite, as seen under the microscope, is shown in Plate 22b; it is the eutectoid which has been formed at 732°C, and the presence of pearlite gives hardness and strength to steel.

With cementite alone the steel would be brittle, with ferrite alone it would be soft. Pearlite combines the good properties of both constituents, though a steel consisting of pearlite alone would be too hard for structural uses. The most widely used steel, containing 0·2 to 0·3 per cent carbon, when in the slowly cooled condition, has a structure of about one-third pearlite and two-thirds ferrite.

Pearlite contains about 0·8 per cent carbon and therefore a steel containing that amount of carbon is completely pearlitic when it has been slowly cooled (illustration *d*). A low-carbon steel has only a small amount of pearlite, while in a steel of over 0·8 per cent carbon the structure consists of pearlite plus surplus cementite (illustration *e*), and such steels possess great hardness and strength, but low ductility. By referring to *Fig. 47* the structure of other steels can be pictured; thus a slowly cooled steel with 0·6 per cent carbon is composed of about two-thirds pearlite and one-third ferrite (illustration *c*).

THE HEAT-TREATMENT OF STEEL

The few simple tests described below may be enlightening. Two good-quality steel knitting needles, a coal fire or gas ring, a bowl of water, a piece of sandpaper, and a pair of pliers are needed. (Steel knitting needles are 25 cm long and are magnetic, and are not to be confused with the now more common anodized aluminium needles which are over 30 cm long and which unfortunately would melt during the following experiments.)

Experiment 1

Take one steel knitting needle and bend it slightly to feel how tough and springy it is. Now hold the needle in the flames, using the pliers, the bowl of water being close at hand. When it is bright red-hot, dip the end of the needle as quickly as possible into the bowl of water. The needle should still be red-hot as it is being quenched. When the needle is cold, try to bend the quenched end. It is hard and brittle and will snap off.

Experiment 2

Take the second needle and heat it until it is red-hot; maintain it at this temperature for about a quarter of a minute. Then withdraw it very slowly, so that it cools gradually. If you now test this end (which you have just 'annealed') it will bend, like a piece of soft wire, and furthermore it will remain bent.

Experiment 3

Heat the needle, which has been softened, to bright red heat, and quench rapidly so that it will be hard and brittle, as in Experiment 1. Clean the needle with sandpaper and hold it above the flames so that it is warmed until a straw colour develops on it; it must not reach even dull red heat. When the tint appears take the needle away and let it cool down. If you bend the end of the needle, which you have 'tempered', you will find that the end so treated is tough and springy.

Thus three heat-treatment processes have been performed on a domestic scale, processes which are being continually carried out, under rather closer control, in factories all over the world. The processes are summarized in *Table 14*.

Table 14

Treatment of needle	Name of process	Resulting condition of steel
Heated to red heat (850°C) and quenched from that temperature	Hardening by quenching	Hard and brittle
Heated to red heat and slowly cooled	Annealing	Soft and not springy
Hardened by heating to red heat and quenching, then warmed to about 250°C, then cooled in air	Hardening and tempering	Tough and springy

The striking variation of properties of steel as obtained in these experiments is associated primarily with its carbon content (in this case 0·7 per cent), and secondly with the rate of cooling from bright red heat.

When slowly cooled as in Experiment 2, the microstructure of the steel of the knitting needle is similar to that shown in *Fig. 47c* and its structure consists of pearlite with some ferrite. The atomic lattice of the steel, when heated to bright red heat, changes to the face-centred cubic arrangement and the carbon is taken into solid solution. This would give the simple microstructure of austenite as shown in *Fig. 47a*. From *Fig. 47* it will be seen that *on slowly heating*, the change in structure in a 0·7 per cent carbon steel starts at 700°C (see line AEB) and is complete at about 730°C (see line CED). The red-hot steel of the knitting needle, when quenched in water as in Experiment 1, cools so rapidly that the carbon atoms have no time in which to come out of solid solution to form cementite, which is one constituent of pearlite. Because of the enforced presence of these carbon atoms the iron atoms can revert only to a distorted form of the body-centred cubic arrangement. This severe distortion is responsible for the great hardness and brittleness produced as a result of Experiment 1. Seen under the microscope this steel in the quenched condition shows a type of structure which is illustrated in Plate 22c and which is called 'martensite' after Adolf Martens, a nineteenth-century German metallurgist.

Few steels are used in the extremely hard martensitic condition and the re-heating operation of tempering is carried out to reduce the brittleness of the steel and yet still to retain much of the hardness, as was demonstrated in Experiment 3.

If the quenched steel is tempered at a temperature as low as 250°C its hardness is only slightly reduced but if tempering is carried out at a higher temperature the reduction of hardness is greater. In practice, steels are not often tempered between 250° and 500°C because of a tendency in that temperature range to become embrittled, a condition known as 'blue brittleness'.

When martensite is tempered by heating it to some temperature below 700°C, particles of cementite begin to form, and in the high tempering temperature range of about 500° to 700°C the globules of cementite attain sufficient size to be seen under the microscope. The structure shown in Plate 22d is of a steel first hardened and then tempered at 300°C; at this temperature the cementite particles have not become visible.

Two important points are to be noted in connection with the quenching and tempering of steel:

(1) To get a fully hardened structure the steel should be heated to a temperature above that represented on the line CED on *Fig. 47* and then rapidly cooled. It will be seen, therefore, that the appropriate temperature before quenching depends on the carbon content of the steel concerned.

(2) Tempering does not restore the pearlitic structure. Before that can be done the steel has to be heated to a temperature above that represented by the line CED and then slowly cooled.

So far we have considered two extremes – very rapid and very slow cooling, as illustrated by Experiments 1 and 2 on page 162. In the industrial treatment of steels, intermediate rates of cooling from high temperatures are also used. One treatment, which is known as 'normalizing', consists of heating a steel to red heat (above the CED line) and cooling it in air. This gives a 'normal' ferrite–pearlite structure similar to that shown in *Fig. 47*, though the pearlite is usually finer.

As in other metal-working operations, a certain amount of mystery has grown around the subject of heat-treatment and quenching liquids and even the origin of the water used for quenching steels. Sheffield water has been credited with unique properties; indeed, one hard-headed Yorkshireman once exported it in kegs to Japan at a fair price. It is, however, probable that the tradition arose from the excellence of the Sheffield craftsmanship and the presence, close at hand, of an excellent deposit of millstone grit used for making grinding wheels.

Today water is rarely used as a quenching medium in the heat-treatment of steel; this is partly because the severe quenching effect which it causes tends to crack large blocks of steel and partly because many modern steels intended for heat-treatment contain alloying elements which dispense with the need for very rapid quenching. Such alloy steels are quenched in whale oil or proprietary oils. *Table 16* on page 175 shows the hardening and tempering temperatures for some typical alloy steels. For comparison the temperatures for two carbon steels are also shown.

CASE-HARDENING

When mild steel or certain alloy steels are heated above 900°C, in contact with carbon, that element is absorbed at the surface of the steel. The depth of the carbon enrichment depends on the time and temperature of the treatment; this process is known as 'case-carburizing'. The mechanism of the process is caused by carbon monoxide being formed; this decomposes to carbon atoms which diffuse into the steel.

A mild steel carburized at about 925°C for eight hours has a depth of outside 'case' of about a millimetre, and a carbon content of about 0·9 per cent at and near the surface. This is capable of being hardened while the inside retains the toughness characteristic of low-carbon steel. When a case-carburized steel has been heat-treated it is said to have been 'case-hardened'. Armour plate and parts of rifles, typewriters, and automobile engines are a few of the multitude of case-hardened steel components.

In the simplest method of case-hardening the steel articles are packed in boxes containing charcoal and other substances; they are then heated gradually to about 925°C, and maintained at that temperature for several hours. The steel is quenched in oil, which is much preferable to water, since too rapid quenching may lead to cracking or peeling of the case.

A thinner carbon case can be applied by immersing the steel in a bath of molten sodium cyanide. The amount of cyanide depends on the work to be treated, varying from 45 per cent for case-hardening thin layers to 75 per cent for thick pieces of steel which are to be given a thick case. Other salts such as sodium or barium chlorides, and sodium carbonate, are added to lower the melting point of the mixture.

Another method of hardening the surface of steels is known as 'nitriding'. Special alloy steels containing aluminium and other elements are heated in ammonia at about 500°C; the gas decomposes to a certain extent into hydrogen and nitrogen; the latter combines with the iron, forming particles of iron nitride which impart great hardness to the surface of the steel. Nitrided steels are used for duties where resistance to wear is important, as, for example, cylinder liners in internal combustion engines, dies, and parts of guns. Nitriding is a more expensive process than case-hardening and is more difficult to control, but nitrided steels do not distort during treatment as case-hardened steels are liable to do, because of the lower temperature of the nitriding process. Combinations of these techniques have been developed and are used to case-harden steels, either by gas carburizing or nitriding methods.

During the last twenty years, several other hardening processes have been developed. In Tufftriding, the parts to be treated – crankshafts, camshafts, gears and sintered components – are pre-heated before being immersed in a titanium-lined pot containing molten potassium cyanide and cyanate at 570°C. Carbon and nitrogen are liberated from the salts in the presence of ferrous materials. Carbides formed on the surface of treated components act as nuclei which promote a tough compound layer of iron carbides and nitrides.

The somewhat similar Sulfinuz process uses a molten salt mixture of sodium cyanide, cyanate, carbonate and sulphide. An iron container promotes the breakdown of sulphides to form an iron sulphide layer which gives additional resistance to seizure and wear to the Sulfinuz-treated steel. There have been other more recently patented processes including 'Sulf B.T.', an electrolytic process which from a molten salt mixture produces a surface layer of iron sulphide. Operating temperatures of 190°C make it an ideal process for tempered steel which would soften if heated at higher temperatures.

12

CAST IRON
AND ALLOY STEELS

Pig iron from the blast furnace is generally converted into steel. However, another ferrous alloy, namely cast iron, is made from pig iron by remelting it in cupolas; these look like small blast furnaces though they melt the iron and do not produce it from ore. The manufacture of all sorts of engineering and structural components in cast iron represents an enormous part of the metal-working field, second only to the steel industry. Over three million tons are produced in Britain each year, which is something like eighty per cent of all metals used in the cast form. The great fluidity of cast iron and its low shrinkage on solidification make possible close tolerances and considerable freedom in design. Cast irons have versatility in properties, including hardness, good machineability, excellent compressive strength, resistance to wear and rigidity. Cast iron is the cheapest metal.

In 1970 the then U.K. Ministry of Technology produced an analysis of the many uses of cast iron. Although lists tend to be dull, this particular report has been reproduced in *Table 15*. Knowing that over three million tons of cast iron are produced each year in Britain, one might guess that motors, lorries and tractors would account for about a third of the tonnage but few would guess that manhole covers or ingot moulds are also responsible for large tonnages. *Table 15* also gives the tonnages of spheroidal graphitic iron and malleable iron, which are discussed later. These figures were reported in tons but can be converted to tonnes by adding 1·6 per cent.

Several foundries in Britain produce over 1,000 tons of iron castings per week consisting of about 500,000 individual castings. Such a foundry would have two cupolas, each capable of melting 25 tons per hour; both may be in operation during the

Table 15. The Uses of Cast Iron, SG Iron and Malleable Iron

Product	Tons produced in U.K. in 1969		
	Grey Cast Iron	SG Iron	Malleable
Pressure pipes and fittings	362,809	96,522	20,552
Rain water pipes and fittings	67,425		1,775
Hot water boilers	67,166		38
Fires and cooking stoves	81,102		46
Baths and sanitary goods	60,307		
Bedsteads, furniture, pianos	4,179		95
Domestic electric appliances	18,867		257
Builders' ironmongery	7,507	1,642	5,146
Manhole covers, gratings, etc.	137,001	452	800
Ships' engines, etc.	35,379	1,197	237
Turbines	4,340	1,021	21
Gas, oil and steam engines	61,027	2,714	91
Pumps and compressors	43,363	1,579	261
Boilers and boilerhouse plant	15,383	371	252
Colliery castings	13,200	526	1,746
Electrical engineering	65,382	2,274	12,989
Switch and fuse boxes	7,805	599	2,410
Chemical and gas plant	17,870	849	372
Textile machinery	54,288	1,456	1,459
Agricultural implements	36,966	779	5,890
Food and drink machinery	10,417	246	114
Machine tools	168,323	3,712	2,437
Printing machinery	30,964	964	385
Tractors	220,535	9,333	25,798
Cars and lorries	614,515	48,814	93,444
Motor cycles	11,896		846
Metal rolls	40,064	4,206	
Sugar and flour mills	8,457		
Locomotives	7,393		
Tunnel segments	35,835		
Permanent-way castings	38,402		7,156
Railway carriages, wagons	30,129		1,365
Other railway parts	6,019	351	758
Cranes	49,852	1,052	3,868
Ingot moulds	591,347	296	
Steel mill equipment	49,669	1,761	718
Valves	63,651	2,295	1,293
Other uses	251,352	9,858	14,248
Totals	3,354,186	194,869	206,867

day and patched during the night ready for the next day's work.

Many varieties of cast iron can be produced, by selection of different pig irons, by variations of the melting conditions in the cupola, and by special alloying additions, but in general the two main classifications are into white cast irons and grey cast irons. Both contain 2·4 to 4·0 per cent of carbon but the difference lies in the condition in which the major portion of the carbon exists in the structure of the metal. In white cast iron all the carbon is present as cementite, and the fracture of such an iron is white. In grey cast iron most of the carbon is present as flakes of graphite, and there is usually a remainder which is in the form of pearlite; the fracture of this type is grey. By far the greatest amount is grey cast iron. White cast iron is rarely used alone, but is an important stage in the manufacture of malleable cast iron, which is discussed later.

Since cementite is intensely hard, white cast iron is hard and durable, though very brittle. Grey cast iron is softer, readily machineable, less brittle, and suitable for sliding surfaces because graphite is soft and is a good lubricant. The rate of cooling to some extent determines whether the iron is white or grey; the more rapid the cooling the greater the tendency to form a white iron. Use is made of this fact in producing large cast-iron rolls, for rolling mills, which have a grey centre but a chilled, and therefore a hard, white iron surface.

Thomas Turner, one of the great metallurgists of the nineteenth century, who later became the first Professor of Metallurgy in the University of Birmingham, published a historic article entitled 'Influence of silicon on the properties of cast iron' in the *Journal* of the Chemical Society in 1885. His research was a landmark in what we described in the first chapter as the art of metal-working growing into the science of metallurgy. Professor Turner proved that the amount of silicon present determines the condition of cast iron. With only 1 per cent silicon the iron tends to be white, while with about 3 per cent silicon, even rapidly cooled irons are grey. Later it was found that the presence of other alloying elements also had an effect on the structure of cast iron; for example, chromium tends to produce a white cast iron and nickel to produce a grey one.

SPHEROIDAL GRAPHITIC CAST IRON

Since 1948 spheroidal graphitic cast iron, better known as SG iron or nodular iron, and in the U.S.A. as ductile iron, has been developed as a constructional material which can compete with steel castings or forgings for many stressed components. It is now produced at the rate of 3,000,000 tons per annum in the western world and is continuing to increase in use by about 15 per cent per annum.

It was discovered that the addition of small amounts – only about 0·04 per cent – of magnesium, or cerium, to molten cast iron would cause the metal when solidified to form the graphite into small nodules instead of flakes. Because the flakes had a large surface area and tended to be sharp at the ends they led to brittleness in the irons. The small nodules of spheroidal graphite, however, allow the cast iron to be much stronger and tougher and more shock-resistant than ordinary grey cast iron.

SG iron is often melted in electric furnaces because these provide greater purity than can be obtained when melting in cupolas.

More than half of all the spheroidal graphitic iron produced is centrifugally cast into pipes in diameters ranging from 50 mm up to 1,000 mm, and used for carrying water and gas at both low and high pressures. The other uses are shown in *Table 15*. One of the interesting markets is the car industry, which is always on the look-out for price savings without loss of quality. Several European car manufacturers have begun to use SG iron for crankshafts, in place of steel forgings. As an example of the reliability of SG iron, the railings of all bridges for forty miles on either side of Manchester on the M6 motorway are made of this particular type of cast iron. When one realizes how important it is that no vehicles crossing over a motorway and inadvertently bumping the parapet of the bridge should cause any damage underneath, one can appreciate that in SG iron we have a material that is tough and dependable.

MALLEABLE CAST IRONS

White cast iron is the basis for the production of malleable iron; it is heat-treated in such a way that the white cementite is broken down into ferrite and graphite of a formation which tends to be nodular in form and gives a black fracture. Such iron is then known as blackheart malleable iron. *Fig. 48* shows the microstructure of the three forms of cast iron. In the ordinary grey cast iron the flakes of graphite are apparent. In the spheroidal graphite, the nodules of graphite can be seen; the form of graphite in blackheart malleable iron consists of nodules of a fuzzier type than those of SG iron. Pearlitic malleable iron was

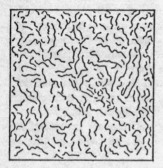

Grey Cast Iron

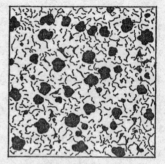

Spheroidal Graphite Iron

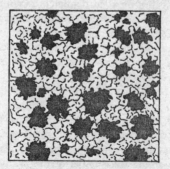

Blackheart Malleable Iron

Fig. 48. Microstructure of three forms of cast iron

developed during the Second World War as a good substitute material for steels. It can be produced by increasing the amount of manganese in white cast iron and heat-treating it. This gives it a pearlitic structure and this malleable iron is harder and stronger than the blackheart malleable iron; consequently most of its present uses are for parts which were originally produced in steels.

Whiteheart malleable iron is produced by packing the castings in iron boxes with iron oxide and heating them for two days at 900–950°C. The malleable iron is then slowly cooled; the inner portion develops a steel-like structure which appears white when fractured.

Malleable iron is an old and, until recently, conservative industry, the process having been discovered two hundred years ago. Perhaps it has the benefits as well as the disadvantages of a process which has been modernized and very much expanded during the last twenty-five years. The industry makes a product which is inexpensive and suited to the needs of modern production requirements because malleable iron is strong, tough and can be machined rapidly. SG iron, the newcomer, is a strong competitor with malleable iron and its arrival undoubtedly stimulated improvements in the older process. As will be seen from *Table 15* their uses overlap. However, malleable iron components are usually less than 60 mm thick, because it would be difficult to produce a very thick white cast iron casting which is the starting point for producing malleable iron. On the other hand, SG iron is not so suitable for use in thin sections.

ALLOY STEELS

A thin piece of steel such as a small tool or needle can be quenched in water and will give uniform properties throughout its section, but if a massive block of steel is heated and quenched the interior of such a block will cool fairly slowly, despite the rapidity of cooling on the outside. Thus there is a gradual decrease of hardness from the outside to the centre of the quenched piece. Furthermore the unequal dimensional behaviour of inside and outside causes internal stresses to be set up, so that a heavy block of high-carbon steel may crack if it is drastically quenched. Such a state of affairs places limitations on the uses of carbon steels, but these difficulties have been overcome, and other advantages conferred, by the development of alloy steels, in which uniform hardness can be produced throughout the whole of a block by a mild quench.

Alloy steels are divided into two types: low alloy steels with under 10 per cent of added elements and high alloy steels with over 10 per cent, usually between 15 and 30 per cent of added elements. During the present century highly stressed components of automobile engines have been of alloy steel. Heavy modern guns could hardly be of anything but alloy steel if their accuracy, trajectory, and range are to be satisfactory. Similar remarks apply to cutting tools for lathes; even bridges are now sometimes of alloy steels. Housewives turn up their noses at ordinary steel knives – only stainless alloy steel cutlery is acceptable.

Alloy steels are used for the following reasons:

(1) To enable the effect of heat-treatment to appear uniformly in large masses of steel without the outer skin behaving differently from the inside, as would be the case if bulky sections of 'straight' carbon steel were treated.

(2) To obtain a given combination of mechanical properties by less drastic heat-treatment than would be necessary in a plain carbon steel.

(3) To enable special qualities to be imparted to the metal such as great strengths, hardness, resistance to wear, springiness or resistance to corrosion.

In order to get the full value of the improvements conferred by the alloying elements, alloy steels are generally used in the heat-treated condition. Carbon is an essential constituent of alloy steels, for without carbon they could not attain their useful mechanical properties; the carbon makes hardening and tempering possible, the alloying elements *modify* the effect of the treatment. One of the earliest alloy steels was that containing nickel, and this was later improved by the addition of chromium, together with adjustment of the amount of nickel. This steel proved to be somewhat temperamental in tempering; it developed a mysterious embrittlement, but a small addition of molybdenum was found to cure this 'temper brittleness'.

Various combinations of strength, hardness, springiness and toughness may be achieved by the selection of times and temperatures of heat-treatment for a particular alloy steel composition. The field is so vast that only a brief mention may be given in the following table.

Table 16

Type of Steel	Approximate Composition	Temperature °C	
		Hardening	Tempering
Die Steel	0·4% Carbon 5% Chromium 1% Vanadium 1·5% Molybdenum	1,000	600–650
Tool Steel	1·5% Carbon 13% Chromium	980	175–420
Air Hardening Tool Steel	1% Carbon 5% Chromium 0·5% Vanadium 1% Molybdenum	970	380–400
Nitriding Steel	0·3% Carbon 0·35% Silicon 1·6% Chromium 1·1% Aluminium	900	500–720
Stainless Steel	0·15% Carbon 16% Chromium 2·5% Nickel	975	650
High-Carbon Steel	1% Carbon	750	180–250
Medium-Carbon Steel	0·6% Carbon	790	180–250

A landmark in alloy steel history was the discovery of manganese steel by Sir Robert Hadfield in 1882. Such a steel, containing about 1 per cent carbon and 13 per cent manganese, can be brought to a high degree of toughness by heating it to 1,000°C and quenching in water. In this state the structure of the steel is austenitic and it is only moderately hard, but any attempt to cut or abrade the surface results in the local formation of hard martensite, so that the steel cannot be machined by ordinary cutting tools. Manganese steel is therefore used for purposes which require intensely hard metal, such as parts of rock-breaking machinery and railway crossings. At the Baker Street junction, ordinary steel rails used to need replacement every nine months, but manganese steel lasted twenty-two years. A modification of this manganese steel is used for military steel helmets for the services, a typical steel containing 1·3 per cent carbon, 1·8 per cent silicon, and 12·9 per cent manganese.

STAINLESS STEELS

In 1913 Harry Brearley of Sheffield was experimenting with alloy steels for gun barrels, and among the samples which he threw aside as being unsuitable was one containing about 14 per cent of chromium. Some months later he noticed that most of the steels had rusted, but the chromium steel remained bright. This led to the development of stainless steels, which possess a very high resistance to corrosion due to a naturally occurring chromium-rich oxide film, invisible, microscopically thin, but protective. The stainless steels, which now often contain other elements besides chromium, are among the most corrosion-resistant alloys available and their use is increasing rapidly. The Scandinavian countries have done a great deal to make stainless steel articles elegant and popular. About 7 per cent of the total Swedish steel production is stainless, compared with about 1 per cent for Britain and 2 per cent for Japan. There are three important categories of stainless steel.

(1) The Martensitic group contains approximately 13 per cent chromium. With carbon at less than 0·15 per cent this type of steel is used for cutlery; when sharper cutting properties are required the carbon is increased to between 0·3 and 0·7 per cent.

(2) The Ferritic group normally contains about 17 per cent

1. The West Driefontein Gold Mine, with the reduction works in the foreground. This is the world's top producing gold mine which in 1970 milled 717,000 tonnes of ore to produce 21,116 kilogrammes of gold.

2. An aerial view of the Kenecott copper mine – the biggest quarry on earth.

3a. Iron ore mining operations at Cerro Bolivar, Venezuela.

3b. Aerial view of the iron ore terminal at B.S.C. Port Talbot Works. (*Courtesy British Steel Corporation*)

4. A modern blast furnace, designed to produce 3,000 tonnes of pig iron per day. (*Courtesy Davy & United Engineering Co. Ltd*)

5a. Concorde. (*Courtesy British Aircraft Corporation*)

5b. Queen Elizabeth 2.

6. The Bubble-raft experiment.

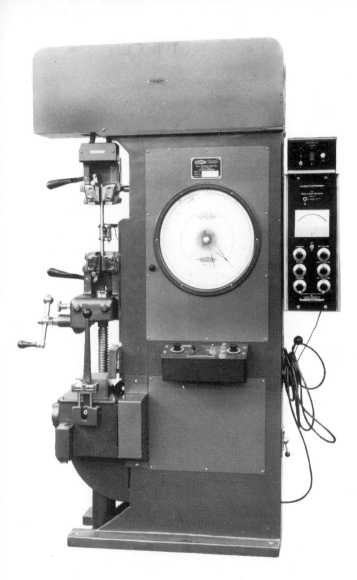

7. A machine for testing the tensile and compressive strength of metals.
(*Courtesy Avery-Denison Ltd*)

8. Casting a *Q.E.2* propeller. The mould in course of manufacture. (*Courtesy Stone Manganese Marine Ltd*)

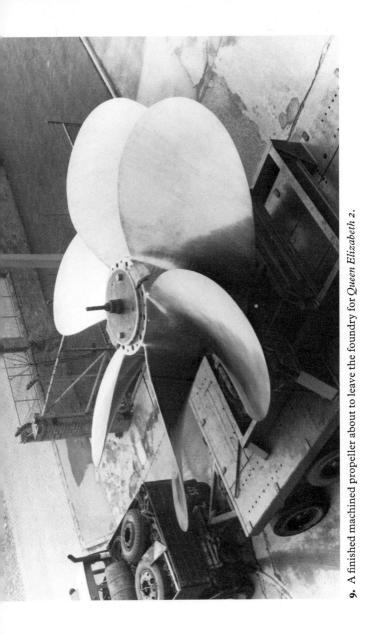

9. A finished machined propeller about to leave the foundry for *Queen Elizabeth 2*.

10a. This Wotan 2,500 tonnes Pressure Diecasting machine was built for the Renault Works in Cleon, France. Dimensionally it is the largest of its kind built in Europe.

10b. Mill with five sets of rolls arranged in tandem, which cold rolls steel strip up to 2 metres wide. It ranks amongst the largest and most powerful of its type in the world and is designed for a high throughput of coils weighing up to 45 tonnes and up to 3 metres diameter. (*Courtesy Davy & United Engineering Co. Ltd*)

11. View from the discharge side of the 3-strand Concast continuous casting machine for steel billets.

12. An 8-strand continuous casting machine for Japan and British Steel Corporation. (*Courtesy Distington Engineering Company*)

13. An L.D. Converter producing steel by the oxygen process. (*Courtesy British Steel Corporation*)

14. 'Eros'. A classic example of aluminium casting, which has stood in Piccadilly Circus since 1893.

15. The 31-storey Alcoa Building at Pittsburg. The exterior was sheathed with aluminium panels formed into the pattern of inverted pyramids.

16. 400kV Suspension Towers on the C.E.G.B.'s Sizewall–Sundon Transmission Line. (*Courtesy Central Electricity Generating Board*)

17. Consumable-electrode vacuum-arc titanium melting furnace. The furnace body is being traversed to the first-melt crucible which contains an electrode of compacted raw material ready for melting. The body of the furnace is cooled by a molten sodium-potassium alloy. (*Courtesy Imperial Metal Industries Ltd*)

18. Rolls-Royce RB.211 three-shaft turbofan. (*Courtesy Rolls-Royce 1971 Ltd*)

19. Cross-channel Hovercraft.

20. Argon-Arc Welding. Part of the Condenser in an air separation column for a steel oxygen plant. (*Courtesy Marston Excelsior Ltd*)

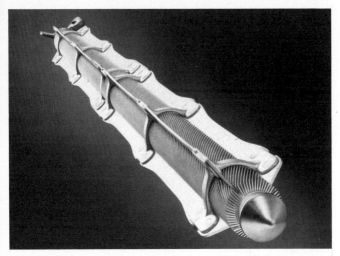

21a. A Magnox can as used at Oldbury Nuclear Power Station. (*Courtesy U.K. Atomic Energy Authority*)

21b. Loading a basket of fuel elements into the charge machine at Chapelcross Nuclear Power Station. (*Courtesy U.K. Atomic Energy Authority*)

22a. Surface of cast antimony, showing dendritic structure.

22b. Pearlite. A steel with about 0·9 per cent of carbon.

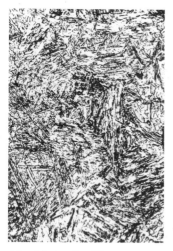

22c. Martensite. The structure of a hardened steel.

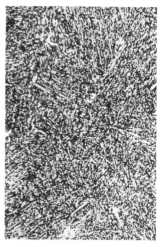

22d. The structure of a steel which has been hardened and tempered.

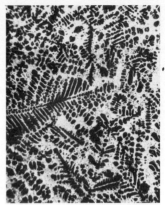

23a. Cuboids of antimony-tin intermetallic compound in a tin-antimony-lead bearing metal.

23b. Alloy of 60 per cent copper with 40 per cent silver, showing dendritic formation.

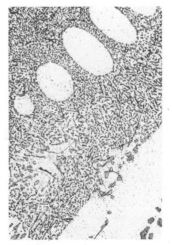

23c. Alloy containing 88·5 per cent aluminium and 11·5 per cent silicon (unmodified).

23d. The same alloy 'modified' by the addition of 0·05 per cent sodium. Compare with 23c.

24a. Dislocations in aluminium which has been cold-worked. Magnification 50,000.

24b. Dislocation lines from particles of aluminium oxide in brass. Magnification 10,000.

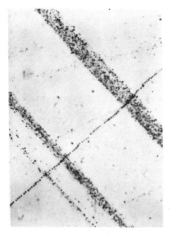

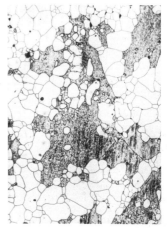

24c. Iron-silicon alloy showing slip-bands. Magnification 2,200.

24d. The same iron-silicon alloy after annealing. The photograph shows dislocations and re-crystallization. Magnification 200.

chromium with about 0·1 per cent carbon and is used for motor car trims. These steels cannot be hardened by heat-treatment, but they can be strengthened by cold working.

(3) The Austenitic group, which is produced in the greatest quantities, contains 18 to 20 per cent chromium with 7 to 12 per cent nickel. Best known is 18 per cent chromium and 8 per cent nickel, used for kitchen sinks. Molybdenum, up to $2\frac{1}{2}$ per cent, is also added for still greater corrosion resistance required, for example, in commercial and industrial sinks, superheater tubing in power stations and the cladding of buildings where atmospheric pollution is high. The steels in the group cannot be hardened by heat-treatment. In contrast to the other two categories, they are non-magnetic.

By adding elements such as 3 per cent copper or aluminium with about 1 per cent titanium to steels containing 17 per cent chromium and 4 to 8 per cent nickel, heat-treatment involving precipitation hardening and/or refrigeration at $-70°C$ can be applied, and strengths of up to 1,500 newtons per sq. mm can be obtained. Such steels are used for aircraft undercarriages and honeycomb stiffened structures for the walls and skinning of missiles and space craft.

In recent years new equipment specially designed for the production and fabrication of stainless steel has been installed to meet the increasing demand. The largest consumers are the domestic equipment and automobile industries; in Britain the manufacture of sinks is the largest single use of stainless steel sheet. In automobiles stainless steel is widely used for external bright trim, windscreen wipers and hub caps. Stainless steel exhaust pipes are, rightly, becoming popular. The brewery, dairy and chemical industries are now using considerable tonnages, and stainless steel razor blades are rapidly displacing ordinary steel.

Stainless steel blades were used before the last war, without much success, because the steel was too soft. The success of the modern blade is due to the presence of a thin layer of the plastic P.T.F.E., which has to be baked on after the blade is sharpened. Conventional razor-blade steel softened when the plastic was being baked on, and was replaced by a suitable stainless steel. Nowadays a martensitic steel containing 13 per cent chromium, 0·6 per cent manganese, 0·4 per cent silicon and 0·67 per cent carbon is used. The utmost technical control is necessary in the manufacture of this material, including even distribution of the martensitic particles, the pressing and clipping operations, heat-treatment, including a final deep freeze, packing and storage.

Stainless steels have been used extensively for architecture in the United States, and the trend is growing. New buildings in south London and the provinces have included large quantities of stainless steel for curtain walling, decorative panels, and door and window frames. Considerable amounts are used for components in power stations, particularly in the field of nuclear power production. As science advances from guided missiles to supersonic aircraft and flights into outer space, further use will be made of stainless steels because they provide such an excellent combination of resistance to heat and corrosion and maintenance of high strength at elevated temperatures.

HIGH-SPEED STEELS

One other type of alloy steel, which is important in the history of metallurgy, and indeed of civilization, is the 'high-speed' steel used for cutting metals on lathes and other machine tools. Straight carbon tool steels are still used occasionally in machining, but fast rates of machining cause such a tool to heat so much that it over-tempers itself and softens. Before A.D. 1900 a cutting speed of 10 metres per minute was considered good for ordinary carbon steel tools; great astonishment was caused at the Paris Exhibition in that year when the Bethlehem Steel Corporation of America exhibited cutting-tool steels, containing tungsten and chromium, that would cut for hours at 50 metres per minute and would continue to cut for some time even when the speed was so much increased that the tip of the tool became red-hot. The new type of steel was developed by Fred W. Taylor and Maunsel White, and later they recommended the use of a steel containing about 0·7 per cent carbon, 18 per cent tungsten, and 4 to 6 per cent chromium. This remarkable stability at high temperatures was largely due to the presence of tungsten carbide, and some chromium carbide.

Using such high-speed steels, metal parts can be machined many times faster than by the use of plain carbon steel tools. In addition to iron, carbon, tungsten, and chromium, high speed steels may contain vanadium and cobalt. These tungsten steels are used for dies in the processes of hot brass stamping and extrusion. Since the discovery of tungsten steel, other tool materials have been developed, for example, molybdenum steels. But one of the best metallic cutting materials turned out to be tungsten carbide bonded with about 10 per cent cobalt. In fact this material is all alloying elements and contains no iron.

NEW DEVELOPMENTS IN ALLOY STEELS

The consumption of alloy steel is going up steadily and efforts have been made to economize in the scarce and expensive elements which are essential constituents of these steels. Sometimes the addition of minute amounts of new elements gives an effect equivalent to that obtained by much larger amounts of other alloying metals. Recent discoveries have led to the addition of small amounts of boron to increase the hardenability of steels of low alloy content. One steel containing 0·15 per cent carbon, 0·40 per cent molybdenum and 0·003 per cent of boron has a high strength and good weldability.

The high-speed steels which are used for cutting tools illustrate the conservation of alloying elements, which became necessary during the 1939 war. It was found that high-speed steels could be made with less alloying content than was formerly considered essential. Steels containing 3 per cent each of tungsten, molybdenum, and vanadium were introduced in Germany; other steels, developed in Britain and the U.S.A., contained 0·8 per cent carbon, 6 per cent tungsten, and 4 per cent molybdenum. These alloying contents are to be compared with 14 to 22 per cent of tungsten which was the normal pre-war range of composition for a high-speed tool steel. Raising the vanadium content to as much as 4 to 5 per cent has proved successful in some high-speed steels.

SUPER ALLOY STEELS

Stringent control of manufacture and the demand for optimum mechanical properties have led to the development of 'super alloy steels' which are completely free from non-metallic inclusions. Such inclusions may originate from refractories with which the steel is in contact in melting and casting; they are potential sources of failure when severe stresses are encountered, as in modern aircraft engines. Among improved methods of handling is the consumable-arc melting technique, which was devised for molybdenum and which has subsequently enjoyed universal use for titanium (Plate 17). A similar technique is now being adopted for melting super steels. Such furnaces can melt up to 8,000 kg.

Vacuum melting is being used increasingly for improving the quality of steels. This technique helps to prevent the formation of non-metallic inclusions and it eliminates deleterious gases. In the older and still popular process the steel is melted in an evacuated vessel. The process is now being extended to larger tonnages by enclosing the ladle into which the steel is cast within a vacuum chamber.

Another method of extracting the last traces of inclusions is by electroslag melting. This was originally developed in the U.S.A. mainly for welding thick sections of steel, but later the U.S.S.R. extended the idea to refine steels by remelting steel electrodes continuously under a flux, the energy being supplied by an electric current across the slag pool.

13
ALUMINIUM

Over a century ago a new metal was being admired at the court of Napoleon III of France. While some nobility at his banquets were served from mere gold plate and silver cutlery, visiting potentates and the favoured few were privileged to use spoons and forks made from aluminium. The King of Siam, visiting the French court, was delighted to receive a watch charm made of this fascinating new light metal. In those days aluminium was worth £100 per kilogram; only a small amount had been manufactured, by reducing a mixture of sodium chloride and aluminium chloride with sodium.

In the early 1890s, when Hall and Héroult had only recently invented the electrolytic process for extracting aluminium, the metal was still an expensive curiosity. It was deemed a 'precious metal' and exorbitant transport charges were accordingly made when aluminium was carried by rail. Nowadays, in the late 1970s, aluminium in ingot form costs over £550 per tonne and, after taking into account its lightness, it competes effectively with the other metals, even with iron and steel.

In 1939 the annual world output was about 750,000 tons and by 1943 had increased to two million tons. After the war production fell, but then started to rise quite rapidly, so that at present the world's aluminium production is about thirteen million tons per annum; Britain uses about 450,000 tons. It is the second most important metal and in good years its rate of growth is about 10 per cent per annum. The industry is dynamic and has done a great deal to develop new uses of the metal, often with bold and imaginative pioneering. The use of aluminium in building, ships, and automobile cylinder blocks was only made possible by a vast amount of design, research and development by the aluminium producers. Since the 1950s there has been over-capacity which has led to competition within the industry and with other metals, sometimes causing a rather unsatisfactory profitability, particularly in the wrought side of the industry. Several properties account for the importance of aluminium.

(1) It is a light metal, with a density about a third that of steel or brass.

(2) Aluminium has a higher resistance to atmospheric corrosion than many other metals, owing to the protection conferred by the thin but tenacious film of aluminium oxide which forms on its surface.

(3) Aluminium conducts electricity well. For the same cross-section area of wire the metal has about two-thirds the electrical conductivity of copper. But aluminium is a better conductor on a weight-for-weight basis, because its density is only a third that of copper.

(4) Aluminium conducts heat well.

(5) Aluminium forms high-strength alloys in conjunction with other elements. Some of these alloys are only one third the weight of mild steel per unit volume and of equal strength.

(6) The large number of useful aluminium alloys can be cast, rolled, extruded, drawn, pressed, riveted, machined and welded without much difficulty.

ALUMINIUM OXIDE AS A PROTECTOR

The famous statue of Eros (Plate 14) illustrates the corrosion resistance of aluminium. The graceful figure was sculptured by Sir Alfred Gilbert and cast in a foundry in London. After over eighty years of exposure to the Piccadilly Circus atmosphere, Eros shows little sign of corrosion. Other examples of the corrosion resistance of aluminium are in food manufacture and storage, in petroleum refining and in the manufacture of nitric acid and explosives where it has replaced fragile earthenware containers.

However aluminium can be a very reactive element and the metallurgist knows that its successful resistance to corrosion depends on the completeness with which the protective layer of aluminium oxide prevents this underlying, lurking activity from coming into play. As soon as aluminium is cut and exposed to the air, aluminium oxide forms on the fresh surface. It occupies the same volume as the aluminium it has replaced; consequently the oxide is neither crumbled nor stretched and seals the surface of the metal so that further oxidation is prevented. Other metals are not so fortunate; corrosion products of iron are soft and crumbly and do not prevent progressive attack.

The natural film of aluminium oxide is only a few millionths of a millimetre thick. It can be thickened up to between 0·0015 mm and 0·04 mm by an electrolytic process known as 'anodizing'. The aluminium articles are suspended in a vat similar to those used in electroplating, but usually containing sulphuric acid solution; the components to be anodized are at the positive, anode, end of the vat. The action of the electric current releases oxygen from the solution and a tenacious coating of aluminium oxide forms on the surface of the metal. Depending on the operating conditions, anodizing can be adjusted to produce either a film of great hardness or one that is very resistant to corrosion. The anodic film also possesses the ability to absorb dyes, thus enabling the metal to be tinted with attractive and enduring colours. For special purposes a hard anodic film, up to a fifth of a millimetre thick, can be produced.

ALUMINIUM AS ELECTRIC CONDUCTOR

Aluminium was first used as an overhead conductor over fifty years ago in Switzerland. Later its use became more general when it was found possible to reinforce the aluminium conductor by a central core of galvanized steel wire. The resulting increase in strength of the cable made long spans possible without too great a sag. This 'aluminium conductor, steel-reinforced' (A.C.S.R.) is used in practically the whole of the British National Grid (Plate 16). Seven unjointed lengths of A.C.S.R. conductor cables, each 4 kilometres long and weighing 23 tonnes, carry an important link in the Central Electricity Generating Board's super-grid system over the rivers Severn and Wye. The crossing of the two rivers at their confluence involved three long spans, one of them nearly two kilometres in length – the longest in Britain. Each conductor consists of 78 aluminium wires of 2·8 mm diameter, arranged in two layers, around a core of 91 galvanized high-tensile steel wires of the same diameter. The seven 4-kilometre lengths involved stranding 4½ million metres of steel and aluminium wire. The main span, 1,600 metres long, has a sag of 60 metres giving a clearance of 40 metres above the water. The outermost layer of wires is wedge-shaped, giving a smooth surface and minimizing wind resistance, to avoid developing harmonic vibrations.

SOME USES OF PURE ALUMINIUM

Pure aluminium foil, often called 'silver paper', has a multitude of uses; about 40,000 tonnes per annum are made in Britain, with a good possibility of increasing to 50,000 tonnes within a few years. The foil is produced by hot rolling aluminium in seven or eight passes till it has been reduced to about 10 mm thick, after which the metal is cold rolled, the end product being coils of aluminium foil only 0·009 mm thick for wrapping chocolates and cigarettes, or 0·05 mm thick for milk bottle tops, cream and yogurt lids. Foil can be produced down to as thin as 0·0045 mm.

Sixty per cent of all aluminium foil is required by the food packaging industry. Meat and poultry for roasting can be wrapped like a parcel in foil, after being covered with fat; no basting is necessary, the foil wrapping retains the juices and prevents cooking odours from escaping. What are now known as convenience foods are mass produced with the aid of foil; the mix is put into individual foil pie dishes which travel on the conveyor to the cooking ovens, and the pies when baked are transported and sold in the same foil containers. About two thousand tonnes per annum are used in Britain for wrapping cigarettes and cosmetics.

The industry is breaking into new fields with considerable enterprise. Foil is replacing copper wire in electrical windings and it is used in dry and electrostatic capacitors, where the aluminium foil is wound with a dielectric material such as paper or polythene. Do-it-yourself car repairs have been made easier with foil kits; exhaust pipes are repaired with a foil bandage impregnated with adhesive. In the pop music scene an embossed and laminated aluminium foil makes attractive record sleeves; the selection entitled Tamla Motown Chartbusters had the best designed record sleeve in 1969. The same type of material is being used in coffee bars to provide patterns and colours which change to the beat of pop music.

Aluminium is a good reflector of heat and there are some applications which take advantage of this property. Storage containers for petrol, milk, and other liquids are coated with crinkled aluminium foil which reflects away the sun's heat instead of absorbing it and so the liquid inside is prevented from becoming unduly heated. Also, for the same reason, aluminium is used in building to maintain constant temperatures in hot and cold weather; the efficiency of aluminium foil for insulation purposes depends upon its being associated with a still air space.

Polished or electro-brightened high-purity aluminium is one of the best materials for reflecting light, far better than ordinary looking-glass and in some respects even more suitable than silver. Aluminium-coated reflectors have two advantages of special interest to astronomers; first, when the front of an astronomical telescope mirror is coated with aluminium it does not tarnish so rapidly as silver; secondly, aluminium reflects ultra-violet light better than silver.

The use of an aluminium mirror coat is now standard practice for all large reflecting telescopes, including the 200-inch Hale telescope at Palomar mountain. This famous telescope is situated in a dry climatic zone, in California; it is re-aluminized about once every five years – a major operation with such a large mirror – and is carefully washed every few months. Mirrors in wetter climates deteriorate much more rapidly; for example the 100-inch mirror of the Isaac Newton telescope, situated near the coast in Sussex, loses its reflectivity within only a few months, due to salt in the air. It is possible to protect aluminized mirrors by evaporating a very thin layer of transparent fused silica on to their surfaces. However, such protective coatings can be made to transmit only a fairly small range of colours and they have not yet found wide astronomical application.

ALUMINIUM CASTING ALLOYS

There is a wide range of alloys available for casting, and a variety of properties which determine the alloy selected for a particular use. Silicon, copper and magnesium are used as alloying elements, either alone or in combination. Often small amounts of other elements are added, including manganese, zinc, titanium and nickel. Most of the alloys have a permitted allowance of about one per cent of iron. All of them can be cast in sand moulds and nearly all can be gravity diecast; a limited number are suitable for pressure diecasting.

The major part of the 200,000 tons of aluminium alloys cast in Britain each year are selected from the 'LM' (light metal) alloys specified by the British Standards Institution, the list being revised every few years to take advantage of new developments. The available alloys in the LM range total over twenty, but of these, eight alloys account for about 85 per cent of the total; *Table 17* shows these alloys. There are other aluminium casting alloys; some are patented compositions including those developed for uses in jet aircraft.

The alloy LM6, containing 10 to 13 per cent of silicon, shows an interesting metallurgical phenomenon known as 'modification'. As normally cast in a sand mould the alloy has a coarse structure, shown in Plate 23c; this is accompanied by low strength and weakness under shock. In 1920 Dr Aladar Paz discovered that a small addition of the compound sodium–potassium–fluoride brought about a change in structure. It was found that other compounds of sodium, or about 0·05 per cent of the metal itself, added to molten aluminium–silicon alloy, modify the structure obtained on solidification (Plate 23d). The alloy treated under these conditions is stronger and tougher than the unmodified alloy and is widely used in its sand cast and gravity diecast forms. Such an alloy has an excellent com-

Table 17. Some Important Aluminium Casting Alloys

Alloy Specification BS. 1490	Main alloying constituents Percentage (single figures are maxima)						Approximate percentage of British Consumption 1975 (%)
	Copper	Silicon	Iron	Nickel	Magnesium	Zinc	
LM2	0·7–2·5	9·0–11·5	1·0	0·5	0·3	2·0	12
LM4	2–4	4–6	0·8	0·3	0·15	0·5	15
LM6	0·1	10–13	0·6	0·1	0·1	0·1	17
LM9	0·1	10–13	0·6	0·1	0·2–0·6	0·1	1
LM13*	0·7–1·5	10–12	1·0	1·5	0·8–1·5	0·5	9
LM24	3–4	7·5–9·5	1·3	0·5	0·3	3·0	17
LM25	0·1	6·5–7·5	0·5	0·1	0·2–0·45	0·1	7
LM27	1·5–2·5	6–8	0·8	0·3	0·3	1·0	5

(LM27 also contains 0·2–0·6% manganese)

*LM13 is a piston alloy. Various producers have specifications based on LM13, while not necessarily adhering to that composition bination of good strength, ductility, resistance to corrosion and good castability. A foundry foreman we once knew was heard to remark, 'it runs like milk'.

Two alloys, LM2 and LM24, contain about 10 per cent of silicon and some copper. These can be produced from scrap and are therefore comparatively cheap; they are more suitable for rapid machining than LM6 and this property is particularly important in the mass production industries where fast machining and long life of cutting tools are necessary. All over the world, these aluminium–silicon–copper alloys are pressure diecast in vast tonnages.

In 1961 an enterprising development made it possible to deliver large amounts of molten aluminium alloy to foundries. Insulated containers holding $2\frac{1}{2}$ tonnes of molten aluminium are taken by lorry, two or three at a time, to the foundry. The containers are set down in a position from which the metal can be distributed, and 'empties' are removed and the user is saved the expense of melting the metal. When one considers the traffic hold-ups that occur in big cities, it may seem a bold plan, but in Britain a two- or three-hour journey from alloy manufacturer to foundry is a practical proposition. The largest 'run' on record happened in U.S.A. A load of molten alloy was taken from the Kaiser plant in Ravenswood, West Virginia, to the Chrysler foundry at Kokomo, Indiana, normally a nine-hour journey. When the lorry arrived, the foundry was unable to accept the metal so the lorry turned round and returned to the aluminium works – a round trip of eighteen hours; on arrival the metal was still molten.

WROUGHT ALUMINIUM ALLOYS

As wrought metal products include sheet, wire, forgings, extrusions and tubes, this group of alloys cannot be classified so simply as those used for castings. Each type of shaping process is covered by an individual specification and the alloys are represented by numbers in each group. Code letters indicate the form of the product and the type of heat-treatment that is given.

The modern wrought alloys based on aluminium with between 1 to 7 per cent magnesium and up to 1 per cent manganese are used on account of their combination of high strength and resistance to corrosion. These aluminium–magnesium alloys, which develop further strength as a result of cold working, are used in the construction of all types of sea-going vessels. The strongest wrought aluminium alloys are made by adding up to $5\frac{1}{2}$ per cent of zinc, with copper, magnesium and manganese. Such alloys, wrought and then heat-treated, have a strength of up to 600 newtons per sq. mm and are used for aircraft fuselages. Special precautions are taken to prevent stress-corrosion (page 273) and to ensure that no severe and weakening notches are included in the design and manufacture of components.

Although tinned steel cans are still used more extensively than aluminium, the easy-open tops of beer and beverage cans require a large amount of aluminium. An alloy with $4\frac{1}{2}$ per cent magnesium is rolled in strip form to 0·35 mm thick, and then V-grooved to 0·27 mm thick, so that the pull-off tab will tear the strip along the groove only.

In recent years the production of all-aluminium cans has grown rapidly in the U.S.A. In Britain, Metal Box Ltd, Reads Ltd and Nacanco Company (a subsidiary of the National Can Corporation) are developing plants with a total capacity of over 800 million aluminium cans in 1977. These are produced by a 'drawing and ironing' process in only two parts – a body and a top. A disc of aluminium is stamped into a cup shape, the walls then being further pressed or 'ironed'. The technique is suited to a soft material like aluminium but there are pilot plants in the U.S.A. and the U.K. for steel. The canning industry has fabulous productivity and the makers of tinned steel cans are responding to the challenge from aluminium with new developments and higher productivity.

Tinned steel bodies with aluminium tops combine the three metals; these cans are used as ferrous scrap, where the small percentages of tin and aluminium are not harmful, but they are not suitable as a source of scrap aluminium. From the point of view of scrap reclamation, there is much to be said for the all-aluminium can. In the U.S.A. the aluminium producers have obtained publicity by exploiting the idea of keeping America tidy by re-cycling used aluminium cans.

AGE-HARDENING

In the early part of the twentieth century, Dr Alfred Wilm, a German research metallurgist, was investigating the effect of additions of small quantities of copper and other metals to aluminium, in the hope of improving the strength of cartridge cases required by the Prussian Government. He tried various combinations and different forms of heat-treatment until, in the autumn of 1909, a revolutionary discovery was made, more or less by accident.

An aluminium alloy containing 3·5 per cent copper and 0·5 per cent magnesium was heated and then quenched in water and finally tested on a Friday. The results were not particularly impressive. Early in the next week, some doubt was expressed about the accuracy of the tests, so the same pieces of the alloy were tried again. To Wilm's surprise the hardness and strength were much higher than the values already obtained; this led him to perform a series of experiments on the effect of storing the alloy for different periods after heating and quenching. The strength gradually increased to a maximum in four or five days, and the phenomenon became known as 'age-hardening'. In 1909 Wilm gave to the Durener Metallwerke at Duren sole rights to work his patents; hence the name 'Duralumin'.*

The Zeppelin engineers of sixty years ago realized that the design of airships might be revolutionized by the use of this light, age-hardened alloy. The rolling of the alloy into sheets and strips presented a number of problems, but intense efforts were made so that most of the obstacles were overcome, and during the First World War large amounts of age-hardened aluminium alloy, strip, sheet, girders, rivets, and other parts were used, first for Zeppelins and afterwards for other types of aircraft.

* Although 'Duralumin' has been used as a family name for a series of alloys, it is now a trade name for aluminium alloys made by one particular company. In this book, however, we use the word in its original meaning.

Metallurgists all over the world attempted to explain the mechanism of age-hardening, but it was not until some years after its discovery that the problem was even partially solved, although of course hardening by quenching and ageing was used extensively in the meantime. Metallurgical detective work in revealing the causes of age-hardening was slow because the instruments then available, including the metallurgical microscope, were not sufficiently sensitive to follow the minute changes involved in the internal structure of the alloy. A discussion of the mechanism of age-hardening is given later in the chapter.

Before 1914 the National Physical Laboratory in Britain had commenced an investigation to see whether age-hardening could be induced in other light alloys and particularly in any alloys which would remain strong at the temperature at which an aircraft piston has to work. This led to the discovery of 'Y-alloy' containing 4 per cent copper, 2 per cent nickel, and $1\frac{1}{2}$ per cent magnesium. The strength of this alloy, now called L35, can be increased by 50 per cent by age-hardening and it can be heat-treated in the cast or wrought conditions. It subsequently provided the basis of a number of special aluminium alloys, such as RR58 containing 2 per cent copper, $1\frac{1}{2}$ per cent magnesium, $1\frac{1}{2}$ per cent iron and 1 per cent nickel. This is used throughout the structure and surface cladding of Concorde, amounting to 70 tonnes in each aircraft.

One complex alloy, used for forged pistons in many British aircraft, is composed of aluminium with about 2 per cent copper and 1 per cent each of nickel, silicon, magnesium, and iron, plus about 0·1 per cent titanium. After an ageing heat-treatment the tensile strength of this alloy is about 450 newtons per sq. mm.

THE MECHANISM OF AGE-HARDENING

Although age-hardening has been exploited by metallurgists and aircraft designers since the 1911–20 decade, it was a long time before the mechanism of the process could be explained convincingly. Indeed, for many years it was not certain which of the impurities or added metals were responsible for age-hardening.

Partly because, even when viewed under the microscope, there was no visible difference in the structure of the treated alloy before and after ageing, the cause remained obscure, and it was not until the electron microscope began to be available that it was possible to explain the mechanism of age-hardening. It is proposed to give a brief outline of the theory of age-hardening as it applies to an alloy of aluminium with about 4 per cent of copper.

When the solid alloy is at a temperature of 500°C, the entire 4 per cent of copper enters into solid solution in the aluminium (see page 73). At room temperature, however, aluminium can hold less than half of one per cent copper in solid solution. When this alloy is quenched in water from 500°C down to room temperature, the copper does not immediately come out of solid solution. This may be compared with the behaviour of some kinds of home-made jams in which sugar crystallizes out, apparently of its own accord, when the jam is left to stand for a long time. The reason is that when the jam is hot it will hold more sugar in solution than when cold; the sugar does not crystallize immediately the jam is cold but becomes apparent several months afterwards.

During the lapse of some days after the aluminium–copper alloy has been quenched, the copper atoms, which can no longer be held in solid solution, are forced to move or diffuse among the aluminium atoms to form minute areas containing higher amounts of copper than the average. These areas ultimately form the intermetallic compound $CuAl_2$ (page 75). They are too small to be seen under the optical microscope but evidence of their existence has been obtained by the use of X-rays and with the electron microscope.

This describes what takes place inside the alloy when age-hardening occurs but why the local aggregations of $CuAl_2$ cause this was controversial for more than fifty years after Wilm's discovery. Gradually it has been realized that the hardness and strength of metals is related to resistance to slip of the crystals and increases in lattice strain. The researches on 'dislocations' described on page 85 have all added to the evidence that hardening is caused by formation of areas or zones of copper and aluminium atoms, which hinder further slip and distortion by preventing the free movement of dislocations (Plate 24a).

PRECIPITATION TREATMENT

The previous discussion referred to an aluminium–copper alloy in which the intermetallic compound $CuAl_2$ is precipitated at room temperature and thus increases the hardness of the alloy. The hardening can be hastened and intensified by heating the quenched alloy at about 175°C for a few hours. This is known as 'precipitation treatment'. Furthermore if magnesium and silicon are also present, a compound of these two elements, Mg_2Si, is formed which takes part in a similar process of hardening to that occurring in ageing. When the aluminium–copper–magnesium–silicon alloy is quenched and reheated to 175°C, particles of Mg_2Si are precipitated in addition to $CuAl_2$. This alloy then becomes harder than if aged at room temperature, as is shown in *Table 18*.

Table 18

Conditions of aluminium–copper– magnesium–silicon alloy	Brinell hardness (approximate)
1. As quenched	60
2. Quenched and kept at room temperature (i.e. naturally aged)	120
3. Quenched and reheated to 175°C (i.e. precipitation treated)	150

This type of hardening has since been discovered in alloys of the majority of metals in common use, such as magnesium, copper, zinc, tin, lead, and iron. Many of these other alloys will not harden by ageing at room temperature. The process involves, first, raising the alloy to a fairly high temperature at which a proportion of the appropriate alloying element goes into solid solution; this is known as 'solution heat-treatment'. After a rapid quench in water, the alloy is precipitation treated to make the dissolved element come out of solution in very minute agglomerations of such a size that maximum hardness is reached.

Alloys which age at room temperature are classed as age-hardening alloys, while those requiring precipitation at higher temperatures are called precipitation-hardening alloys. It might be noted that the essential difference between the precipitation treatment of such alloys and the heat-treatment of steel is that steels attain their maximum hardness by quenching, while tempering usually *reduces* the hardness. The precipitation-treated alloys, as quenched, are comparatively soft, but precipitation treatment *increases* the hardness.

The process of age-hardening aluminium alloys in aircraft factories introduced the refrigerator into metallurgical practice. When 'Duralumin' was first used, a part such as a rivet had to be driven almost immediately after quenching, otherwise the age-hardening change would commence and the alloy would become too hard to work. It was found that if, immediately after quenching, the alloy was stored in a refrigerator at about 15°C below zero, the age-hardening change was slowed down and the rivets could therefore be stored at that low temperature until they were required.

CLAD ALUMINIUM ALLOYS

An aircraft may consist of precipitation-hardened aluminium alloy to the extent of over a third of its weight. Much of this is in the form of thin sheets which are used for covering the wings and fuselage. Unfortunately alloys of this class do not have very good resistance to corrosion, and this makes them liable to lose strength.

To provide the necessary protection and thus retain high mechanical strength, a process was developed for giving aluminium alloy sheet a coating, by rolling a thin layer of pure aluminium on to each side of the sheet, thus giving a 'three-ply' metal (*Fig. 49*). This is such an efficient combination of protective metal with strong alloy that sheets 1·5 mm thick can be exposed for over five years to the continuous action of salt spray without deterioration through corrosion. The thickness of the aluminium coating varies according to the gauge, but as

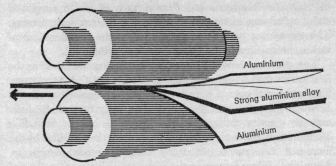

Fig. 49. The cladding of aluminium alloy sheet

a general rule the total thickness of the aluminium, front and back, amounts to 10 per cent of the thickness of the sheet. This is the principal form in which aluminium alloy sheet and strip, in heat-treated condition, is used in aircraft today.

The rolled clad material is heated rapidly to about 500°C and quenched, thus bringing it to the soft condition. It is then beaten, pressed, or otherwise worked to the required shape. After that, the clad metal is aged in a temperature-controlled furnace, when the alloy core of the sheet increases in hardness and strength owing to precipitation-hardening.

USES OF ALUMINIUM ALLOYS

Plate 15 shows an American skyscraper with the outer walling built from aluminium panels. Very many buildings constructed since the war have used aluminium 'curtain wall' structures. Also the metal is used for window frames, doors, roofs and venetian blinds.

Soon after 1945 it was realized that there would be great advantages if aluminium could be used to improve stability and permit larger and higher superstructures of ships. More than half the weight of a superstructure can be saved if aluminium is used instead of steel. At first aluminium was tried for small vessels including lifeboats. Some of the milestones in the development of aluminium in ships include S.S. *United States* in 1952, then *Oriana*, *Canberra* and *France*, each of which employed about 1,000 tons of welded aluminium in their superstructures. The experience with these vessels led to the confident use of welded aluminium in the *Queen Elizabeth 2* (Plate 5b).

The alloy with about $4\frac{1}{2}$ per cent magnesium was used in the four major decks above the steel portion of the *Q.E.2* structure. The additional cost of the thousand tons of aluminium alloy is between £250,000 and £500,000. The basic cost of the liner may be £20–25 million, so the increased cost of including aluminium is about 2 per cent. This is more than justified by the increased revenue obtained by carrying more passengers. Aluminium alloys feature in the Hovercraft (Plate 19), a class of vessel in which weight saving is obviously important. The conception of an all-aluminium passenger ship is still a possibility.

The early nineteen-sixties brought many new uses of aluminium in automobiles, including the spectacular achievement of the diecast aluminium alloy cylinder block. The introduction of the compact car coincided with the change-over to aluminium for many components and in 1961 seven makes of American cars had die-cast cylinder blocks. However, it was still necessary to produce these diecastings with cast-iron cylinder liners and economic considerations caused the projects to be discontinued after a few years. In Europe, where cars are lighter, diecast cylinder blocks are used by Renault and Peugeot. The Russians produce diecast cylinder blocks and the liners are assembled into the blocks in the course of machining.

In America, General Motors overcame the difficulty of inserting liners by using an aluminium silicon alloy containing about 17 per cent of silicon and 4 per cent of copper. Because the amount of silicon is greater than the eutectic composition of about 11 per cent, the alloy is called hypereutectic. This composition has great fluidity when molten, so is suitable for diecasting. In order to ensure solidity General Motors use a modification of the diecast process known as 'Acurad', which involves a secondary injection plunger and a thicker gating system than is normal in pressure diecasting. After the cylinder block has been diecast, heat-treated and machined, an electrochemical treatment is applied to the bore which removes a layer of aluminium about 0·002 millimetre deep. This leaves the hard silicon grains projecting slightly from the bore, to form a wear-resistant surface. The 4-cylinder block is required for the Chevrolet Vega engine. Diecasting machines of 2,500 tons locking power are used and the plant at Massena, near the U.S.–Canadian border, produces over 50,000 tons of diecastings per annum. The production of the cylinder block involves precise control of metal composition, automation of the process, automatic conveyor equipment and temperature control of the die.

In Britain the development of high production sand-casting technology has led to the sand cast aluminium alloy cylinder blocks for the Rover V8, Jaguar V12, Aston Martin V8 and Rolls Royce V8. The small 4-cylinder engine of the Reliant car is an aluminium alloy gravity diecasting. The aluminium alloy cylinder block of the Hillman Imp is produced by a variation of gravity diecasting called low-pressure diecasting.

Aluminium alloys are so light that many structures made with them are stronger than steel but only about half the weight. Although such light alloy structures may cost twice or three times as much as steel, the higher capital cost of an aluminium bus, railway coach, or cabin cruiser may be offset by reduced running charges. In some cases, by the re-design of a structure in aluminium alloy to replace steel, it has been possible to make overall economies because the greater thickness of aluminium has led to savings in stiffener members and hence less labour charges for assembly, outweighing the higher first cost of the material.

A heat-treated alloy containing about 1 per cent silicon and 0·7 per cent magnesium has considerably better corrosion-resistance than most other age-hardening alloys and is used for bus chassis and ship parts; it was also used for the girders of the Dome of Discovery at the 1951 Festival of Britain South Bank Exhibition and in the 65-metre trusses of the huge hangars for Comet aircraft.

Two developments in the jointing of aluminium and its alloys have helped widen the use of aluminium alloys. One is the development of inert-gas-shielded metal arc welding, mentioned on page 286. The other is brought about by using aluminium–silicon alloys containing 5 or 12 per cent silicon. These alloys can be rolled as a coating on to the aluminium alloy, 'cladding' it in the same way as described on page 194. Then the clad strip is formed into an appropriate assembly, for example, the channelled structure of an aero-engine radiator. It is clamped in jigs, preheated in a high-speed air-circulating furnace for a few minutes, and then immersed in a special flux bath kept accurately to temperature. The aluminium–silicon alloy fuses and joins all the abutting pieces together, forming a 'brazed' radiator block. Large assemblies of such corrugated structures for heat exchangers are used in the manufacture of tonnage oxygen for the steel industries of the world.

In 1938 Germany produced 165,000, U.S.A. 130,000, and Canada 65,000 tonnes of aluminium. However Germany's consumption of the metal was over twice that of U.S.A. Apart from Germany there were five major producers in the West – Alcoa, Alcan, Pechiney, Norsk Hydro, and Alusuisse. The crash programme to develop the U.S. industry led to the birth of two more giants – Reynolds and Kaiser.

Since then a large number of other producers have become active. By the mid-seventies the total world production will probably amount to 14–15 million tonnes per annum of which that of the Free World will be about 11·5 million tonnes, but it is doubtful whether that much will be required. The industry has considerable problems. The aluminium producers are probably hoping, maybe praying, for something to turn up akin to the use of aluminium in ships or motor cars, which was a feature of the development of this industry in the nineteen-sixties.

14
COPPER

Five hundred years before the Flood metallic copper was discovered in that part of the world now called Iraq; its attractive colour and the ease with which it could be beaten into useful shapes were valued by the early civilizations. It is believed that about 3500 B.C. copper was accidentally smelted, by a fire that had been made in contact with a copper-containing mineral, thereby reducing the ore to metal. In Cyprus almost pure copper was made; in fact the island got its name from that of the metal. In other parts of the world, where tin and copper ores exist together, bronze was smelted.

Copper and bronze were known in Britain long before the Romans arrived; bronze weapons dating back to 1800 B.C. have been found here. During the Roman occupation of this country they mined and smelted copper ores in Cumberland, Anglesey, Wales and Cornwall. The production of copper continued in Britain and by the early part of the nineteenth century most of the world's somewhat limited requirements of copper was smelted in Swansea.

Until about the middle of the nineteenth century copper was regarded as a rather splendid metal of somewhat limited use, suitable mainly for roofing churches, making bronze bells, small cannons, massive doors and brass hardware. Then a number of developments led to the growth of electrical engineering and telecommunications. As often happens, the large-scale production of a metal coincides with the expansion of man's need for that metal. The following list shows some of the highlights in the growth of electrical engineering, linked with the career of one man who saw the possibilities of copper and who was partly responsible for expanding the mining of copper ores to meet the demand.

1831   Michael Faraday discovered the principle of electro-magnetic induction, which made possible the development of electric motors and dynamos.

1856   A young Irishman, Marcus Daly, left his home in Ballyjamesduff, to seek his fortune in America.

1861   Antonio Pacinotti invented the ring winding system, using copper, which made the electric dynamo a practical proposition. His work was published in an obscure Italian journal, but received little attention.

1862   Daly got a job as foreman of a silver mine in Nevada and from that time became well known as a mining engineer and manager with an uncanny ability to assess the potential values of metal ore deposits.

1866   A telegraph cable was successfully laid across the Atlantic.

1870   Zenobe Gramme re-discovered Pacinotti's invention and the dynamo began to be developed.

1875   Michael Healey, prospecting for silver, staked a claim on a hill in Montana. He remembered a newspaper editorial which had said 'General Grant will encircle Lee's forces and crush them like a giant anaconda'. Healey gave the memorable name to what was later to be called the richest hill on earth.

1876   Alexander Graham Bell transmitted speech by a copper telephone wire.

1878   Thomas Alva Edison produced his incandescent electric lamp.

1879   The working of an electric dynamo was demonstrated at the Berlin Exhibition, where a small electric locomotive pulled three cars containing twenty passengers.

1880   Michael Healey had now staked several silver claims near Anaconda but was needing more capital to expand them.

1881   Marcus Daly, now a prosperous mine manager, met Healey who offered him a share of the Anaconda property. Daly at once began to deepen and develop these silver mines, which often struck irritating outcrops of copper ore, which nobody particularly wanted.

1882   Edison opened the Pearl Street generating station in New York, to supply electricity for 5,000 lights.

1882   Daly discovered rich copper ore at Anaconda and had a hunch that this metal was worth developing.

1882–1884   Daly shipped 37,000 tons of rich copper ore to Swansea for smelting.

1883   Daly began to operate a copper smelter in Anaconda. This was in full swing by 1884.

Thus a man of energy and vision was in the right place at the right time and within a few years the rapid growth of the use of electricity caused a tremendous surge in the production of copper. Apart from the discoveries in U.S.A., copper ores were mined in other parts of the world. Although at first many of these ores were imported and smelted in South Wales, eventually the more economical way of smelting near the mine was developed, as Daly had done, and Britain lost her proud position of centre of the copper industry. During the twentieth century extensive copper mines have been opened up and smelting plants erected in Australasia, Canada, Chile, Japan, Peru, Philippines, South Africa, U.S.A., U.S.S.R., Zaire and Zambia.

Copper has a higher conductivity of heat and electricity than any other substance except silver. The pure metal is ductile and malleable and can be rolled into strips less than 0·25 mm thick, or made into foil only 0·02 mm thick; drawn into wires less than 0·02 mm in diameter; pressed, forged, beaten or spun into complicated shapes without cracking. Such ductility is also possessed by several copper alloys, notably brass.

Copper and its alloys have attractive appearance and colour, ranging from red in the pure metal to ochre, gold, yellow or white in its various alloys. They can be cast with ease and with beautiful results, as exemplified by the bronze castings that have been made during the last three or four thousand years. Copper and most of its alloys can be joined by such processes as soldering, brazing and welding. It is resistant to many forms of corrosion and when, with the lapse of time, copper roofs become tarnished by the atmosphere, the result is an impressive green patina.

Over seven million tonnes of copper are produced each year, third in tonnage only to iron and aluminium. However, its production is not increasing, except in years of great prosperity, and it must be admitted that not so many new and exciting uses have been found for copper, compared with the growth of aluminium. This is partly due to economic factors. Copper is generally found in weak ores containing only about one per cent metal. Thus high energy and labour costs are involved in converting the ore into copper. The ores are mined either in parts of the world where labour is already highly paid, or rapidly increasing in earnings.

Strikes in copper mining are rather prevalent and the price of copper tends to escalate and decline according to the current supply and demand situation. In 1969 copper rose to £700 per tonne, declined in 1970, and started to increase again in 1971, reaching over £500 in March. However, by November 1971 the price had fallen again to under £400, after which it rose once more. In 1973 all commodities increased in price; copper reached £1400 per tonne. Since then it fell to about £600, then increased towards £800 by late 1976.

COPPER AS CONDUCTOR OF ELECTRICITY AND HEAT

Over a hundred and fifty years ago William Cooke and Sir Charles Wheatstone put a copper telegraph wire on a section of the London and North Western Railway, between Euston and Chalk Farm. Soon afterwards attempts were made to lay telegraph wires under the sea and in 1850 the brothers Jacob and Watkins Brett laid a single copper wire, covered with gutta-percha, across the English Channel. The wire broke after being in operation for only one day, but it was replaced a year later with an armoured cable, having a four-wire stranded copper conductor which proved to be satisfactory. Then in 1857 came the first attempts to connect a cable across the Atlantic, culminating in success in 1866.

About half the output of copper is used as the pure metal, for electrical requirements. For long-span overhead cables high conductivity of electricity is needed, but the metal must also be strong, to support its own considerable weight and to withstand additional stresses due to the effect of wind and the accumulation of ice. The strength can be improved by alloying, but any increase of strength thus obtained is offset by lowering of conductivity; for example, the addition of 10 per cent aluminium doubles the strength of copper but reduces the conductivity to about one-sixth. However, one element, cadmium, allows a useful compromise to be effected. When about 0·8 per cent of cadmium is alloyed with copper the conductivity is reduced to only about nine-tenths that of pure copper and at the same time the strength of the cold-worked alloy is much increased. This cadmium–copper alloy is used for contact wires in railway electrification and for the impressive cable spans over the River Thames at Dagenham.

Because the price of copper has been increasing faster than that of aluminium and because of the price fluctuations referred to on page 202, aluminium is taking the place of copper for overhead conductors and underground cables. Aluminium cables ready for use are about half the cost of copper cables with the same current-carrying capacity. On the other hand other electrical components require different properties, such as the space or cross-section which the conductors occupy, ease of jointing and the ability to stand high temperatures. In all these properties copper is better than aluminium so it is preferred for bus-bars, switchgear, transformers, electrical generators and electric motors.

The use of copper containing small additions of cadmium, silver or phosphorus in motor-car radiators illustrates the suitability of such materials for conducting heat. Furthermore, copper has good soldering and brazing properties, and resists corrosion. The complicated honeycombs used in a modern radiator are made only with considerable difficulty; other metals or alloys which might be preferable on account of their greater strength or lightness have not been used commercially yet, because of greater difficulty of joining, inferior corrosion-resistance, and lower heat-conductivity.

COPPER AND OXYGEN

Even if copper were initially produced free from oxygen, the operation of casting the molten metal would normally result in a pick-up of oxygen from the atmosphere, forming cuprous oxide. If the oxygen content is above about 0·2 per cent, this causes the copper to be brittle. On the other hand in the absence of oxygen, hydrogen is absorbed by the molten metal from moisture or fuel gases; when the metal solidifies, the hydrogen, not being so soluble in the solid metal as in the liquid, is liberated and leads to a characteristic porous structure. Moreover, if both hydrogen and oxygen are present together in the molten copper there is risk of unsoundness, due to the liberation of steam, formed by interaction of the hydrogen and oxygen, during the solidification of the metal. In practice, oxygen is usually kept between 0·025 and 0·05 per cent.

The most important grade of copper is high-conductivity which is 99·9 per cent pure and is used in generators, bus-bars, power lines, switches, electric motors, welding machines and electric traction. The next important tonnage usage of copper is in its deoxidized form, containing between about 0·015 and 0·05 per cent of phosphorus. Such copper is easy to work and to weld, and hence is used for domestic water pipes, boilers, heat exchangers, automobile radiators, petrol and oil lines, and many other welded fabrications.

Oxygen-free copper is produced from cathode copper, by a process which prevents the absorption of oxygen and hydrogen from the environment. It is used for parts of radar and other electronic apparatus. Oxygen-free copper is the basis for special alloys containing elements such as zirconium, chromium and magnesium, used for components of switchgear which need a combination of good conductivity and high strength at elevated temperatures.

BRASS

The best-known copper alloys are the brasses, which are composed of copper and zinc. The art of brass casting in

Britain dates from 1693, when John Lofting, a London merchant, was granted a patent for the casting of thimbles, which previously were imported from Holland. Lofting, with three men and three boys, was soon casting 140 gross of thimbles per week.

Brasses containing less than 36 per cent of zinc are ductile when cold and can be worked into complex shapes without the necessity of frequent annealing. A cartridge case, as used by N.A.T.O. for a 7·62 mm bullet, is an example of the use of a brass containing 30 per cent zinc, familiarly known as 70/30, or cartridge, brass. A strip about 3·2 mm thick is cut into circular blanks about 30 mm in diameter; then each blank is pressed into the shape of a shallow cup. By a series of further pressing operations the cup is pushed and squeezed through successively smaller and smaller holes in a steel die resulting in the walls of the cup being elongated until eventually a tube with a relatively thick base and thin walls is obtained. Further operations are carried out to the base to form the recess for the detonating cap and ensure that the cartridge fits only one type of breech. The rim of the cartridge is also softened by annealing so that it may be bent in to clip the bullet.

The brasses which contain above 36 per cent of zinc are harder and stronger than those containing less than 36 per cent. This fact has already been noted on pages 81–3, where it is shown that with up to 36 per cent zinc, the alloy consists of an alpha solid solution. Between 36 and 42 per cent, another solid solution, beta, is also present and these alloys are called alpha-beta brasses; among these, 60/40 brass is the best-known example. Although such brasses are less workable at room temperature, their plasticity is increased at high temperatures; they are usually shaped by hot-rolling, extruding, hot-stamping, casting, or diecasting.

While the alpha brasses are often straight alloys of copper and zinc, other alloying metals including aluminium, iron, tin and manganese are added to alpha–beta brasses, and their strengthening effects are in the order in which they are named, that of aluminium being greatest.

Very often lead is introduced into brass to improve its machineability. When such a 'free-cutting' brass is machined on a lathe, the metal, which contains particles of lead, does not cling to the cutting tool in long spirals but breaks off in small chips, so a leaded brass can be machined at a much higher speed than would be possible in the absence of lead. For obtaining maximum cutting speed, about 2·5 to 4·5 per cent lead is included in the alloy, which is then used for making screw-threaded products. The presence of such an amount of lead in the brass brings about some deterioration in the mechanical properties and tends to make hot-stamping difficult, so, where brass has to be shaped by hot-stamping and then machined rapidly, only about 1 to 2·5 per cent of lead is introduced.

BRONZE

Strictly speaking, bronze is an alloy of copper with tin, but the word has come to signify rather a 'superior' material as compared with common brass. For example silicon bronze and aluminium bronze contain no tin, while manganese bronze contains only a small amount. The 'bronze' which is most familiar is the copper alloy containing $\frac{1}{2}$ per cent tin and $2\frac{1}{2}$ per cent zinc, from which old pennies, halfpennies and farthings were made, as well as the decimal two, one and half pence.

Before 1672 the humbler coinage was of silver but for economic reasons copper farthings and halfpennies were introduced. Later pennies and twopences were made, in Boulton and Watt's foundry in Birmingham; the pennies were so large and heavy that they were nicknamed 'cartwheels'.

The use of bronze instead of copper for coinage was one of the consequences of the French Revolution. The revolutionary atheists destroyed churches and tried to find a market for the bronze church bells. It was found that by adding an equal weight of copper to the bronze from the bells, a good coinage alloy was formed. During the years after the Revolution more and more copper was added to the bell bronze until, half a century afterwards, by trial and error, an alloy of 95 per cent copper, 4 per cent tin and 1 per cent zinc was adopted in France and then by many other countries, including Britain in 1860. The changes in the composition of British bronze coinage are shown opposite.

Table 19. Composition of British Bronze Coins

Date	Copper per cent	Tin per cent	Zinc per cent
1860–1923	95	4	1
1923–1942	95½	3	1½
1942–1945	97	½	2½
1945–1959	95½	3	1½
1959 to date (including decimal coinage)	97	½	2½

A well-known modification of copper–tin bronze is phosphor bronze. One type of this alloy contains 4·5 to 6 per cent of tin and less than 0·3 per cent of phosphorus, which then exists in solid solution in the bronze. Such a phosphor bronze is very suitable for springs, and electrical contacting mechanisms where the alloy's resilience, non-magnetic properties, and freedom from corrosion maintain regular working.

When over 0·3 per cent phosphorus is present, the surplus separates as a hard constituent, Cu_3P; this type of phosphor bronze, which usually contains about 10 per cent of tin, is extensively used in the form of castings. It is harder than the first type and makes a good bearing-material and is also used for components which endure heavy compressive loads, such as gun mounting rings, parts of moving bridges and turntables, and rolling-mill bearings.

The alloys known as gunmetals are also copper-base alloys containing tin, and this type of alloy was known from early times, when it was used for cannons. Because gunmetal may be cast with ease, it still enjoys a wide reputation, though not for making guns. One of the classical alloys, Admiralty gunmetal, containing 88 per cent copper, 10 per cent tin, and 2 per cent zinc, has been used for marine purposes, for pump bodies, and in high-pressure steam plants. The leaded gunmetals headed by the ubiquitous 85–5–5–5 alloy (85 per cent of copper with 5 per cent each of tin, zinc, and lead), dominate the sand-founding industry for general-purpose and pressure-tight castings.

LEADED BRONZE BEARINGS

The leaded bronzes are an interesting series of alloys which are becoming of increasing importance for bearings. When very high speeds and pressures are encountered, as in a modern aero-engine, the lubricant tends to be squeezed away from the bearing surface. The function of lead in a bearing bronze is to act as a sort of metallic lubricant when the oil film breaks down. In bearings for modern engines the leaded bronze consists of copper with about 30 per cent lead, with additions of 5 per cent or more of other elements such as zinc, tin, or nickel. The copper provides high thermal conductivity, which assists in avoiding overheating, while the presence of the minor additions improves the lead distribution and increases the mechanical strength of the bearing. These alloys are sometimes used for grinding-machine bearings; the lead acts as an absorbent, in which minute particles of grit can safely embed themselves and thereby reduce wear on the remainder of the bearing surface.

ALUMINIUM BRONZES

The aluminium bronzes are copper–aluminium alloys to which other elements such as iron and nickel may be added. They have a golden colour, they are strong and resist corrosion well, especially marine corrosion. The mechanical properties of an aluminium bronze depend on its aluminium content, as was shown on page 66.

Products requiring extensive cold working, such as tube, sheet, strip and wire, are generally made in the softer and more ductile aluminium bronzes, with up to 8 per cent aluminium. Plates of up to 4 tons for heat exchanger and condenser tube-plates replace brass, since the aluminium bronzes in general have better mechanical properties, together with resistance to corrosion and erosion, both in sea water and polluted estuary water. Much aluminium bronze is used in shipbuilding for both naval and mercantile marine, since it can be used as a substitute for steel, both above and below the water line where strength combined with non-magnetic properties are required, such as in underwater detection gear.

The aluminium bronzes containing from 8 to 11 per cent of aluminium are used in the sand-cast, gravity diecast, and hot worked forms. Typical examples are gear selector forks for automobiles, valve gear inserts, gears, locking rings, masonry fixings and dies for both compression and injection plastic mouldings. During recent years these alloys have become popular for ships' propellers and now 80 per cent are in aluminium bronze, including the nickel–manganese–aluminium bronze supplied for the *Queen Elizabeth 2*.

OTHER ALLOYS

Although copper has a very high electrical conductivity it is possible, by alloying, to decrease the conductivity so much that some copper alloys are used because of their great resistance to the passage of electric current. One such copper alloy contains 13 per cent of manganese and 2 per cent of aluminium. These alloys also have the useful property that their electrical resistance does not change with variation in temperature; they are used for underfloor heating and for resistances to control the speeds of electric motors.

Other copper alloys are referred to elsewhere, copper–nickel alloys on pages 63 and 230, 'nickel silvers' on page 232, and beryllium copper on page 240. Partly because copper forms alloys with so many other metals but also because it has been used for thousands of years, very many copper alloys – over 500 of them – have been in use, ranging from the modern temper-hardening copper–chromium alloys to the tin bronzes, which are as old as the art of metallurgy.

15

FOUR COMMON METALS – TIN, ZINC, LEAD AND NICKEL

TIN

Tin is a soft weak metal, slightly less dense than iron and with a low melting point (232°C). It is expensive, costing about £5,000 per tonne, one of the reasons for this being that tin-bearing ores have a very low metal content, often less than one per cent. The high price of tin causes it to be used for components where the maximum advantage can be taken of the metal's corrosion-resistance coupled with minimum thickness and weight. In that context it is valuable throughout the whole spectrum of industry, from cars to cans, from heavy engineering to micro-circuits. Yet the man in the street probably thinks tin is cheap; one often hears the remark 'it's only a bit of tin'. The explanation is that by far the largest use of this metal is as tinplate, which is thin steel sheet, coated with a much thinner layer of tin for protective purposes.

Tinplate is an excellent combination of metals because it has the strength and rigidity of steel, plus the attractive appearance, corrosion-resistance and good solderability of the tin coating. If steel alone were used for food storage the metal would rust, while tin alone would be too soft and much too expensive.

The story of tinplate's progress from an interesting invention to a mass production industry began with Napoleon Bonaparte.

Napoleon was concerned with the problem of feeding his troops who, as every schoolboy knows, marched on their stomachs and whose efficient movement over great distances in good fighting order provided one of the shock tactics which enabled Napoleon to win so many battles. In 1795 he offered a prize of 12,000 francs for a method of keeping food fresh for long periods. The prize winner was Nicholas Appert of Paris, who discovered that when foodstuffs were boiled in glass bottles and immediately sealed, they would keep for several months. This led to other processes being tried, and among them was a method, patented by an Englishman, Peter Durand, of enclosing food in containers of iron, coated with a thin film of tin. This was the first time that tinning had been used for the preservation of food, although the art of coating iron with tin had been practised on a large scale in Bohemia in the thirteenth century when tinned iron was used for decorative articles and for parts of armour.

Tinned cans of a hundred and fifty years ago were made so solidly that opening them was more of a job for a safe-breaker than for a housewife. Some early tins bore the instruction 'cut round top with a hammer and chisel'. Today, by contrast, the emphasis is on easy-opening and the latest development in tin cans is the tear-off end both for food and drink containers. Some problems concerning this apparently essential feature of our civilization have been discussed on page 189.

In the nineteenth century the manufacture of tinplate was practically a British monopoly, until 1891, when the U.S.A. imposed tariff protection to this manufacture. By 1912 the U.K. and the U.S.A. were making three-quarters of a million tons each, but during the 1914–18 war, British production was halved and that of the U.S.A. doubled.

In 1939 total world production had risen to about $3\frac{1}{2}$ million tons; there was a decline during the war years, because many tin-mining areas were in countries affected by the conflict with Japan. After the war, Britain and the U.S.A. were still major producers, but other countries began to develop tinplate manufacture. By 1974 the total world production amounted to 14,780,000 tonnes, of which the U.S.A. contributed 5,200,000 and Britain 1,130,000 tonnes. This British tinplate involved the use of 7,000 tonnes of tin (a proportion of only 0·6 per cent). In that year about 450 million aerosol cans were produced and about 10,000 million cans for all other purposes.

Tinplate is made with mild steel, containing about a tenth of

one per cent of carbon. Steel ingots are hot-rolled down to slabs
and thence into long strips about a metre wide, 3 mm thick and
up to 1,000 metres long. Coils of this steel, weighing about
12 tonnes, are brought from the hot-rolling mill, pickled in acid
to remove scale, and prepared for cold-rolling, to reduce the
thickness from 3 mm down to the finished gauge, which nowadays
can be as little as 0·17 mm. The cold-rolling is done by a 5-strand
high-speed tandem mill (page 107). The maximum speed at exit
is about 1,500 metres per minute. After removing all traces of
rolling lubricant, the steel is annealed by heating in a non-
oxidizing atmosphere. Then it is given a final, very light rolling
treatment without lubrication, which confers the required
mechanical properties and surface finish.

Over 95 per cent of world production of tinplate is manufactured by electrodeposition in continuous automatic plants
with annual capacities of up to 200,000 tonnes of finished tinplate, able to handle the strip at speeds of over 300 metres per
minute. Electrodeposition has almost entirely superseded the
long-used hot-dip process of immersing sheets in molten tin
because of advantages in production speed, control of thickness
and uniformity of the coating, and ability to produce a thinner
deposit. It has a special advantage, that different thicknesses of
tin may be applied to the two faces of the strip. The coating
thicknesses in common use are in the range 0·0004 to 0·002
millimetre.

In about 60 per cent of tinplate production, the tin is electro-
deposited from a solution containing stannous sulphate, phenol-
sulphonic acid and organic addition agents which ensure a
smooth coherent coating. The steel strip is fed continuously
through what is aptly called a 'vertical serpentine' electrolytic
tinplate line. It is about 100 metres long and at any one time a
length of over 900 metres of strip is travelling up and down in
vertical loops as it is uncoiled, cleaned, pickled, plated, momentarily heated to fuse the tin, passed through a 'passivation
solution', oiled, automatically inspected and re-coiled. At full
speed the steel strip takes only about three minutes in passage
from one end of the line to the other.

The brief fusion of the tin coating, known as 'flow brightening' gives brilliance to the coating. The passivation solution, of sodium dichromate, confers improved resistance to oxidation and staining; the oiling process is applied to ease the handling of the tinplate in can-making.

Most of the remainder of the electrolytic tinplate production uses what is called the 'halogen process' in which the tin is deposited from a weak acid solution containing stannous chloride and the fluoride of an alkali metal with an organic addition agent, and the steel passes through the cleaning and plating sections horizontally. One face is plated before the strip is looped back for plating the other face. Otherwise the sequence of operations is as described for the vertical serpentine process.

Tin's low melting point and ready ability to alloy with other metals are reasons for its good solderability, so tin and tin–lead alloys are frequently employed to coat electrical or mechanical components, which are later required to be soldered. About one tenth of the world's output of tin is used in the manufacture of solder; this is usually an alloy of tin and lead, although a little antimony is included when greater strength is required. Solders play their part in telecommunications, the generation and distribution of electricity, the supply of gas and water, in cars and television sets, and electric light bulbs. A vast tonnage is required for the soldering of tin cans, carried out automatically at the rate of up to fifteen cans per second.

Pewter is a tin alloy; there are two compositions in general use, one with 6 per cent antimony and 2 per cent copper; the other contains 4 per cent antimony and 2 per cent copper. Although pewter may be regarded as an old-fashioned metal it is benefiting from an increase in popularity resulting from the introduction of new methods of manufacturing pewterware and under the stimulus of modern design. Pewter is an easy metal to fashion and many home handymen and women have had their first experience in metal beating with pewter for ornaments and vessels.

Bearing metals contain tin; for example, one bearing metal used in diesel engines and generators contains 90 per cent tin, 7 per cent antimony, and 3 per cent copper. At the other end of the scale the lead-alloy bearing-metals, referred to on page 223, contain 5 to 10 per cent tin; these are used, for example, for the bearings of rolling mills.

An important bearing alloy for motor-car engines is an 80 per cent aluminium 20 per cent tin alloy developed in Britain and now used throughout the world. Bronze and gunmetal, which have already been referred to on pages 206 and 207, are other well-known alloys containing tin. The musical sound of church bells is due to the sonority of bronzes containing tin. Many organ pipes are of tin alloy, the sound being considered much better than from pipes made of zinc or aluminium. Another example of the musical use of bronze is in the manufacture of cymbals, consisting of copper alloyed with 15 to 20 per cent tin, which alloy is worked into the characteristic disc shape. The manufacture of the best cymbals was for years a monopoly of Turkish craftsmen, but is now being achieved in Britain.

A fairly recent development has provided one of the most exciting uses of tin. The Pilkington float glass process involves running molten glass on a bath of pure liquid tin; the contents of the bath amount to anything up to 100 tonnes. Tin was chosen for this purpose because of its combination of properties: the requirements were for a supporting liquid which would have a density greater than glass, be molten below 600°C, but have a high boiling point and have virtually no chemical interaction with glass.

New uses for tin are continually being developed, the most recent being its incursion into ferrous products. Small amounts of tin added to cast iron improve its wear resistance and its ability to be machined. Currently, tin powder is being incorporated into iron powder for the production of sintered iron components which are employed in many sophisticated machines. Tin in all its applications can be said to live up to its reputation as a material of civilization.

ZINC

The camp-fire of our forefathers led to the discovery of many metals. When oxide ores of copper, tin, lead or iron were accidently heated in a fire, the carbon in the burning wood united with the oxygen in the metal compound, leaving a more or less pure metal to be found when the fire had died down. Zinc, however, could not have been discovered easily in this way; at bright red heat the liquid metal boils and so, although metallic zinc may have been formed in the camp fire, it would become vapour which quickly oxidized in contact with the air, forming a cloud of white fumes, easily mistaken for smoke.

Long before zinc was known as a metal, the Romans mixed calamine, which contains zinc carbonate, with copper ores; the smelting of the two materials produced brass. Zinc was not isolated for many hundreds of years after the discovery of brass; it was first made in Sumatra and China, whence it was exported to Europe in the early seventeenth century. England was the first European country to develop the manufacture of zinc; William Champion of Bristol was smelting the metal on a commercial scale in 1740.

Zinc is contained in sulphide ore deposits widely distributed throughout the world. The most important mines are in Canada, U.S.S.R., Australia, U.S.A., Peru, Poland and Japan. Recently discovered major deposits in Eire are now being mined. Zinc is often smelted or electrolytically refined in the countries where the mines are situated but large quantities of ore, usually concentrated by the flotation process (page 34), are also sent for treatment to Belgium, Britain, Germany, Japan and the U.S.A. The total world production in 1975 was 5,500,000 tonnes, of which Britain used 209,000 tonnes.

The original process for the production of zinc was a thermal one, and such methods still account for a third of the metal made. The zinc oxide formed by roasting the sulphide ore was heated to a temperature of about 1,100°C with anthracite or a similar carbonaceous material in banks of small horizontal fireclay retorts. Zinc was formed as a vapour, which was caught as liquid metal in condensers adjoining the furnace; each retort had a daily output of about 30 kg of metal.

In the late 1920s an important modification to the thermal method was developed. In this – the vertical retort process – a briquetted mixture of roasted concentrates and bituminous coal is heated in a large vertical retort made of silicon carbide bricks, in which reduction can proceed continuously. This semi-mechanized plant greatly reduces the labour needed, in comparison with hand-operated retorts, but its initial cost and maintenance are greater. The daily output of a vertical retort is 8 to 9 tonnes of metal. By a subsequent process of fractional distillation, a metal with a zinc content of more than 99·99 per cent can be produced for the manufacture of zinc diecasting alloys and for other purposes.

The electrolytic zinc process, developed during the First World War, now accounts for about three-quarters of world zinc production. Roasted concentrates are dissolved in sulphuric acid; and, after intensive purification of the solution, zinc is deposited electrolytically on aluminium sheets from which it is stripped off, melted, and cast into slabs; the acid is simultaneously regenerated and used repeatedly. The purity of the metal so formed is greater than 99·95 per cent and can be maintained above 99·99 per cent when desired.

The Imperial Smelting Process

It was recognized long ago that blast furnace production of zinc would be more efficient than the vertical retort process but various attempts to design a suitable furnace were unsuccessful. After works experiments which started in the 1940s, Imperial Smelting Corporation, at Avonmouth, announced in 1957 the successful development of a blast furnace for making zinc. This was a promising achievement which involved the simultaneous smelting of zinc and lead, the two metals normally occurring together in nature.

In the Imperial Smelting Process, zinc and lead sulphides are roasted to produce oxides which are charged in the blast furnace with proportioned amounts of coke. Preheated blast air enters the furnace through water-cooled tuyères. Slag and molten lead containing precious metals and copper from the charge are tapped from the bottom of the furnace and separated. Zinc leaves the shaft as a vapour in the furnace gas containing carbon monoxide and carbon dioxide and is shock-cooled in a lead splash condenser, the zinc vapour being absorbed by the lead. Next the hot lead containing zinc in solution is pumped out of the condenser and cooled so that zinc is rejected from the solution and floats on the lead. The zinc layer is tapped off and cast and the cooled lead is recycled to the condenser.

The invention of the lead splash condenser provided the key to the process as its extremely rapid cooling prevented oxidation of the zinc vapour by carbon dioxide which had frustrated earlier researchers. There are twelve Imperial Smelting furnaces, capable of producing about 800,000 tonnes per year in total; progress with this considerable technical achievement was somewhat delayed by environmental problems. Much of the zinc is redistilled to make metal of higher purity.

Uses of Zinc

The main uses of zinc are for protecting steel from rust, for brassmaking (see page 204) and for zinc alloy diecasting, which together take over 80 per cent of the total consumption. Its other uses include rolled zinc for building, battery cans, zinc dust for protective paints and zinc oxide, which is an essential ingredient of rubber mixes, some paints and ceramic goods, for coating electrostatic copying paper and to make various other zinc chemicals.

Zinc sheet was adopted as a roof covering in Europe early in the nineteenth century. Many British railway stations and seaside piers were roofed with zinc and a life of 40 years could be expected if the work was done properly. The roof of Kemble station, in Gloucestershire, has lasted over 100 years. Rolled zinc is also used for photoengraving plates, addressograph plates, organ pipes and trunk linings.

Zinc-Coating Processes

Galvanizing, in which steel is coated with zinc, accounts for a third of the world's use of zinc. A household dustbin is one of the most familiar articles in galvanized ware; other examples are metal windows, electricity pylons, electric railway overhead supporting structures and, in North America, the underbody construction of cars. The coating is applied in a somewhat similar manner to the making of hot-dipped tinplate, mentioned on page 212. Over the last forty years, a development in hot-dip galvanizing has been the large-scale production of continuously galvanized steel strip, which has replaced individually dipped sheets. This material has a ductile coating and will stand the same amount of deformation as the steel itself.

Electro-galvanizing consists of electroplating zinc on the previously cleaned article. Although electro-galvanized coatings have a good appearance, they are used primarily to prevent corrosion. As well as small nuts and bolts, zinc plating is used to protect steel strip by a continuous process. The coating is thinner than that applied by hot dipping and is intended as a base for painting, to give added protection.

Zinc is coated on to metals for protective purposes in several other ways. One method consists in spraying zinc from a 'metallization pistol'; zinc wire or powder is fed into the pistol, where it is melted by an oxy-gas flame and, in atomized form, is blown on to the article by a stream of compressed air, the method being similar to that of a scent spray. This process is used for covering ships' hulls and anchors, bridge steelwork, and large tanks. Metal sprayed coatings of aluminium and zinc give life-long protection to steel structures.

Another zinc coating process is 'Sherardizing', where the steel articles to be coated are placed in a rotating drum, together with zinc powder. The container is then sealed, and heated for several hours at about 375°C, when the zinc diffuses into the steel, giving a fine-grained protective coat of zinc–iron alloy. Small parts such as springs, washers, nuts and bolts are now Sherardized automatically.

Diecasting

A modern car is a travelling exhibition of zinc alloy diecastings. Large American cars have used over two hundred and fifty, with a total weight of up to 100 kg. Many of these diecastings are electroplated with copper, nickel and then chromium, the door handles being perhaps the most familiar examples. The carburettor,* petrol pump, door handles, lamp housings, shock absorber bodies, ventilation grilles and sometimes the radiator grille are examples of the scope of diecasting, which has been

* Some automobile carburettors are diecast in aluminium alloy.

described previously on page 95. *Fig. 50* shows the Stag car, indicating some of the zinc alloy diecastings.

At the other end of the scale, model cars such as the famous 'Matchbox toys' are splendid examples of the accurate reproduction obtained by diecasting. The fully automatic production of these cars is admired throughout the world for the high standard of productivity, helped by superb diemaking and efficient small diecasting machines.

Gramophones and record players provide another outlet for zinc alloy diecastings. Many components are diecast, including the tone arms, some of which are cast with a section thickness of only about half a millimetre. Turntables offer special problems to the diecaster since they must be flat, and balanced to a high degree of accuracy. Modern washing machines use zinc alloy diecastings for lids, doors, control handles, pulleys and functional components including gear cases and the gears themselves.

Unalloyed zinc dissolves iron and steel, so the pure metal would not be suitable for making diecastings. When 4 per cent of aluminium is alloyed with the zinc, this effect is inhibited; furthermore the alloy casts with ease and it is strong. However, these zinc–aluminium alloys are very sensitive to the effect of traces of certain impurities. This is in contrast with most other foundry alloys, where small amounts of impurities are far from being catastrophic. The zinc–aluminium alloys produced from zinc of the quality manufactured in the 1920s contained traces of tin, lead and cadmium, which caused the diecastings to become brittle after a few months; in some cases they disintegrated.

It was a difficult problem because little more than one part of tin in 50,000 would spoil the zinc alloy. If a piece of tin the size of a 1p coin were dropped into a crucible containing 250 kg of molten zinc alloy, the material would be spoilt. Eventually ways of overcoming the difficulty were devised. A small amount of magnesium added to the alloy, about one part in 5,000, was to some extent an antidote to the effect of the impurities. Then the process of making high-purity zinc was considerably improved, the oxide being mixed and compressed with carbon to give briquettes which became the charge for a continuous smelting process. By careful redistillation, a zinc containing less

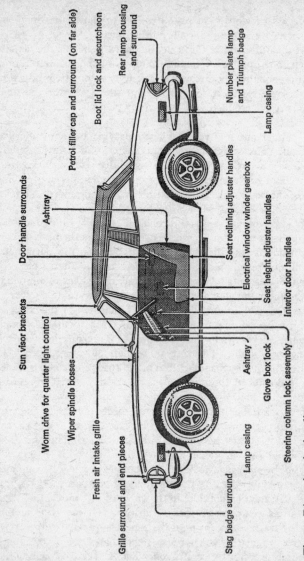

Door handle surrounds

Ashtray

Sun visor brackets

Worm drive for quarter light control

Wiper spindle bosses

Fresh air intake grille

Grille surround and end pieces

Stag badge surround

Lamp casing

Petrol filler cap and surround (on far side)

Boot lid lock and escutcheon

Rear lamp housing and surround

Number plate lamp and Triumph badge

Lamp casing

Seat reclining adjuster handles

Electrical window winder gearbox

Seat height adjuster handles

Interior door handles

Ashtray

Glove box lock

Steering column lock assembly

Fig. 50. Diecastings in zinc alloy

than one part of total impurity in 10,000 parts of zinc was produced. More recently it has been found economical to make the special high-grade zinc electrolytically.

It is this 'super pure' zinc, known as 'four-nines' (over 99·99 per cent zinc) that is used today and the diecast zinc alloys are no longer unreliable. The purity of these alloys is so essential that in recent years diecasters in Britain, America, France, Germany, and Australia have evolved certification schemes under which they guarantee to use nothing but zinc alloy of maximum purity. As a further precaution, each day's production is analysed, and the castings held in bond till each batch has been certified as being of the required purity of composition. The zinc diecasting alloys are generally known in this country as 'Mazak', a name with an American flavour, derived from the initial letters of the words Magnesium, Aluminium, Zinc, and 'Kopper'.* In the U.S.A. the alloys are called 'Zamak'.

LEAD

The Romans' elaborate system of water distribution made it necessary to obtain large amounts of lead; Britain was their principal source of supply. For Roman baths the metal was cast into thick sheets, which were used for lining. Water pipes, in lengths of about 3 metres, were made by bending a lead sheet over a rod so that a tube shape was produced, and the joint was sealed with solder or with lead. The Romans must have realized that lead has good resistance to the attack of water; some of the lead pipes made two thousand years ago have been excavated at Pompeii, Rome and Bath, and found to be in good condition. The Latin word for lead was *plumbum* and the artificers who worked in this metal were called *plumbarii*; some Roman plumbers were women, so that perhaps the term 'plumber's mate' had an even more appropriate meaning then than it has today.

A property which makes lead easy to work is its low melting point (327°C). Molten lead can be cast into shapes from Buddhas

---

* In the early days of the diecasting industry the alloy contained some copper, but this element is now rarely included outside France and Germany.

to bullets, and its ease of casting is one of the reasons why for centuries lead has been associated with the manufacture of printers' type metal; for example, this book has been printed using an alloy containing 74 per cent lead, 16 per cent antimony, and 10 per cent tin.

The combination of ease of melting and low strength made the ancients regard lead as the least noble of metals, and it will be remembered that in *The Merchant of Venice* the casket chosen by Bassanio was of 'base' lead. Yet notwithstanding its low rank in the family of metals, lead was used for coinage in early times because it was easy to cast and because it resisted corrosion.

The world's annual production of lead is about four million tonnes, though only about two-thirds of this is newly smelted metal. Lead is so easy to melt and refine that a large amount of metal can be reclaimed from scrap batteries, sheet, pipes and cable sheathing. Oxides and impurities are removed, the alloy is brought to the required composition, pumped from the furnace and automatically cast into ingots, the processes requiring careful, experienced metallurgical control. Apart from the scrap treatment described above, the main sources are from lead ores in the U.S.A., Canada, U.S.S.R., Peru, Mexico, Australia, Yugoslavia, Ireland, Sweden and Spain. The United Kingdom obtains ore supplies from Australia and Canada, who also export large amounts of metallic lead to this country. Britain's consumption of lead amounts to about 300,000 tonnes per annum, of which about a third is made into anti-knock additions to petrol and various lead compounds required by the paint industry and for filling battery plates.

About 66,000 tonnes of lead, as metal and oxides, are used for making batteries in U.K., while in the U.S.A., their batteries require 500,000 tonnes – over half their total consumption of lead. There are over 17 million motor vehicles in Britain, of which about 13¾ million are cars; all of them use lead-acid batteries. The British car and motor-cycle market requires over 6 million battery units per annum, of which 4½ million are replacements and the rest for new vehicles. Motor-car battery plates consist of grids, made of a lead alloy containing antimony, automatically gravity diecast at very high rates of production; lead oxide pastes are pressed into these grids. Larger versions of the car battery are required in the 60,000 electric vehicles and 80,000 battery-driven electric trucks used in factories.

Lead has been used in building work since early Egyptian times. Nearer home there are many lead roofs which, like those on Westminster Abbey and St Paul's, have lasted for several centuries. Lead is the most widely used of all materials for flashings and weatherings on chimneys and the tops of parapet walls. In spite of the competition from plastics and copper, lead pipe is still required in building, due to the ease with which it can be manipulated and fitted in confined spaces. A recent use of lead has been for sound insulation in public buildings. Lead sheet can be bonded to plywood and chipboard and these laminates make effective partitions with good sound insulation.

Lead is essential for the sheathing of underground telephone and power cables. The sheath prevents the access of moisture to the insulating materials and conductors. The softness and ductility of lead enables it to be applied to very long lengths of cable without harming the insulating materials.

The third important outlet for metallic lead is in the form of alloys used in engineering and manufacturing industries. Lead will not dissolve completely in some other liquid metals; for example molten lead and aluminium will not mix. Alloys of lead and copper, containing between 40 and 95 per cent lead, are not fully miscible in the liquid state except at temperatures in excess of 1,000°C. Leaded bronze (page 208) is an excellent bearing metal, made by rapidly solidifying a suspension of lead particles in copper. However lead does form useful alloys with metals of similar low strength and melting point. About 14,000 tonnes per annum of lead–tin alloy solders are made in Britain. Apart from battery plates, mentioned above, the lead–antimony alloys are used in small arms ammunition and for ornamental castings.

An important use of lead in engineering is in the making of bearing metals. Such lead-base alloys are not quite so satis-factory as tin-base alloys or leaded bronze for severe conditions of high speed service, but for medium speeds and loads they are suitable and their low cost is an advantage. The principal type of lead alloy bearing metal contains lead, antimony, and tin.

Lead can be rolled into sheets with light rolls and it can easily be extruded into rods, pipes, and tube containers. The low strength of lead makes it unsuitable for engineering where considerable stresses are encountered, and it is not surprising that while work was progressing with the strong alloy steels, aluminium, magnesium, nickel, and copper alloys, little attention was paid to the possibility of improving the strength of lead, which has only a fraction of the strength of steel. Nevertheless, there are good reasons why attempts should be made to improve the mechanical properties of any lead components which are subjected to stress in service. When the water in ordinary lead pipes freezes, the expansion which occurs produces enough stress to burst the pipe after two or three freezings. Several interesting alloys were developed in this country which gave improved properties to lead pipes. The addition of less than a tenth of one per cent tellurium was so effective that a pipe would stand five freezings before it burst. For cable sheathing there is growing interest in an alloy of lead with antimony and arsenic.

In the past, lead has been thought of as somewhat unglamorous, though capable of doing a great deal of metallurgical donkey-work. However, to mix our metaphors, lead, like other dark horses, often brings surprises. Many modern developments of lead and its alloys have been enterprising and sophisticated.

In nuclear research establishments and in hospitals where X-rays, radium and radioactive isotopes are involved in examination and treatment, lead is used in the form of interlocking bricks made of the metal alloyed with 4 per cent of antimony. Radiography and hospital treatment rooms are constructed with sheet lead attached to other materials, such as plywood. Altogether about 15,000 tonnes of lead are used in Britain for radiation shielding. A rather interesting use is made of finely powdered lead dispersed in polythene where shielding is required against both gamma rays and neutrons at the same time. This metal/plastic combination can be fabricated and machined into rods, slabs, sheets and complicated shapes, and can be injection moulded as though it were pure plastic.

Many new composite materials are being produced by the lead industry. The metal is used in combination with plywood and other building materials and with foam rubbers and polystyrene in order to provide sound insulation. Dispersion-strengthened lead is a recent development in which the relatively soft and weak metal is hardened and strengthened by the introduction of a small amount of other materials – a common and very effective dispersant is lead oxide. Lead is also used as a coating for steel; this material finds use in containers which require the strength of steel but cannot tolerate the way in which it rusts if unprotected. Lead coatings on steel are used to facilitate soldering components together and the fins of motor car heaters are now made of an extremely thin steel strip coated with lead.

NICKEL

The name 'nickel' originated in Germany; sometimes the early copper miners in the Harz mountains found ores which were contaminated, and could not be reduced into workable copper. They attributed this to the influence of 'Old Nick' and one of the contaminated ores was nicknamed 'Kupfer-nickel', which might be translated as 'Devil's copper'. It was later found that this difficulty was due to the presence of a compound of arsenic with another element, which was henceforth named 'nickel'.

Today the nickel–iron sulphide ores occurring in Canada, U.S.A., New Caledonia, Australia, U.S.S.R., Cuba and South Africa provide the major part of the world's nickel. To augment the rather limited supplies of these ores, vigorous efforts are being made to win the metal from nickel laterites, which are claylike deposits containing nickel oxide, occurring to a far greater extent than the sulphide ores. However the lateritic ores are usually found in the tropics, remote from civilization. They contain much moisture which has to be removed; they are very low in nickel content and they contain other elements which are difficult to separate. Thus every lateritic deposit is an individual riddle – a particular combination of geological and chemical factors. Nevertheless these ores will be needed to an ever-increasing extent to supply the world's hunger for nickel.

When nickel was first isolated by Axel Cronstedt in 1751 it was brittle and could not be used alone, although it could be included in many excellent alloys. However, some fifty years later it was shown that the brittleness was due to the presence of impurities, chiefly carbon and sulphur. The addition of manganese lessened the embrittling effects of these impurities and nickel was then found to be a ductile and workable metal. Today, magnesium is added to nickel, either alone or in conjunction with manganese. In recent years very pure nickel, completely free from impurities, has been produced; it is ductile without the addition of manganese and magnesium.

The original Orford process for the separation of the nickel and copper sulphides in the ore was developed by William Gibb. In this process the mixture of metallic sulphides is melted with sodium nitrate; when cooled and allowed to solidify it forms two separate layers, the top one containing the copper and the bottom the nickel. Recently this process has been modified so that on slow cooling the two phases can be separated easily by grinding and flotation. The nickel sulphide is then roasted, to form nickel oxide; this is reduced to metallic nickel which is purified either by an electrolytic process or, in Britain, by the Mond process. In 1889 Ludwig Mond and Alfred Langer discovered that carbon monoxide gas combines with nickel when heated to 60°C, to form nickel carbonyl gas; when this is heated to a higher temperature it breaks down to give pure nickel and carbon monoxide. In the process developed from this discovery the nickel forms into pellets, and the carbon monoxide gas is recirculated for further use. The metal is marketed as electrolytic cathode in powder or pellet forms, as ferro-nickel in pellet and ingot, and as oxides.

Demands for nickel have increased steadily over the years and, thanks to intelligent marketing and well-produced literature, the uses of nickel have become increasingly diverse. The free world consumption of nickel in 1974 was 580,000 tonnes though, with the world depression, it fell to 440,000 tonnes in 1975. About 43 per cent is used in stainless steels, 20 per cent in nickel-rich non-ferrous alloys, 11 per cent in electro-plating, 10 per cent in nickel alloy steels, 9 per cent in iron and steel castings, and the remainder in copper alloys, permanent magnets, chemicals and a wide range of other applications.

Uses of Pure Nickel

Nickel is a metal similar to iron in some of its properties. It has a melting point of 1,454°C, nearly as high as that of iron; it has slightly lower strength and hardness and is magnetic, though to a lesser degree than iron. In contrast to iron, however, nickel is strongly resistant to corrosion and is used for this purpose in industry. For example, it is not corroded by alkalis or chlorine, and is used for the construction of plant making these materials.

Its ease of working, resistance to corrosion and relatively high strength at elevated temperatures led to its use in the construction of radio valves, cathode-ray tubes, and other electronic devices. The low electrical resistivity and resistance to scaling at high temperatures have resulted in the use of nickel, alloyed with small percentages of manganese and silicon, for automobile sparking-plug electrodes.

The best known use of nickel is in electroplating, the nickel forming a continuous, ductile, corrosion-resisting layer which is then covered with a thin film of chromium. Although well-known plated articles, ranging from car door handles to domestic fittings, are often loosely described as 'chromium-plated', the resistance to corrosion depends on the layer of nickel; the chromium merely provides a thin bright hard coating over the nickel. The chromium, if applied alone, would be brittle and of little value in preventing corrosive attack.

Articles to be nickel plated are suspended in a bath containing a solution of nickel sulphate, with small amounts of other salts added. Anodes of pure nickel are also suspended in the bath and electric current flows from these anodes to the articles to be plated. The development of 'bright nickel' solutions has reduced the need for the amount of polishing which used to be done both before and after plating. The nickel-plated goods are taken out of the bath, rinsed, and have a fine polished appearance; they are suitable for chromium plating without further polishing.

Nickel Steels and Cast Irons

In 1889 James Riley of Glasgow published a now historic paper on alloys formed by the addition of nickel to steel. This report was brought to the attention of the U.S. Navy, who took up the idea and tested armour plate of nickel steel, finding it superior to ordinary steel plate. From that time onwards nickel steels have been used for components in automobiles and other vehicles, aircraft and other equipment where high strength and harden-ability are required. Nickel steels have already been mentioned in chapter 12 and so have the stainless chromium–nickel steels for engineering and for corrosion-resisting purposes.

In 1925 a fresh development occurred in the ferrous alloys, when nickel cast iron began to be used, the nickel giving increased hardness, strength, and resistance to wear, together with good machining properties. The amount of nickel added is usually below 3 per cent, although higher amounts are added to give exceptional hardness or resistance to corrosion. Crushers and ball mills for cement, gold ores, and coal are some of the uses of these very hard, nickel cast irons.

Low-Expansion Alloys

An alloy composed of iron with 36 per cent nickel has the property of expanding only to a very small extent when changes in its temperature occur. The fact that metals expand when heated will be familiar to those who have observed the expansion gap left between lengths of railway lines. If a tape made of aluminium were laid from Marble Arch to Dover it would expand about 3 metres when the temperature rose by 1°C; if the tape were of steel it would expand $1\frac{1}{2}$ metres; but a similar tape of the 36 per cent nickel–iron alloy would expand only 8 centimetres. The discoverer of this alloy, first known as 'Invar', was Dr Charles E. Guillaume of the International Bureau of Weights and Measures in Paris. From the date of its discovery in 1896, Invar has been used for accurate measuring tapes and in chronometers, where its use prevents inaccuracy being caused by temperature changes.

By using other nickel–iron alloys the thermal expansion of the metal can be selected; for example, an alloy of iron with 46 per cent of nickel has the same expansion as platinum, which in turn is similar to that of glass. Previous to the discovery of this alloy, platinum lead-wires were needed for electric bulbs, otherwise the glass would have cracked owing to the unequal expansion of the glass and metal with heat. The nickel–iron alloy can be used instead of platinum and it is much cheaper. Another nickel–iron alloy, containing 78·5 per cent nickel, is unusually susceptible to magnetism of very low intensities and was embodied in the transatlantic cable, where its application increased the speed of transmission fivefold. Nickel features in some of the powerful permanent magnet alloys which have been developed in recent years, such as the alloy of nickel, iron, and aluminium known as 'Alcomax' or that of aluminium, nickel, and cobalt known as 'Alnico'.

Nickel–Chromium Alloys

The alloy of 80 per cent nickel and 20 per cent chromium possesses a high resistance to the passage of electricity, coupled with resistance to corrosion and scaling at bright red heat. Although nowadays it is taken for granted that electric fire and furnace resistances will not keep 'burning out' it is remarkable that this alloy can work for so long at very high temperatures without deterioration. The highest grades of nickel–chromium resistance wires have the 80/20 composition but lower grades are made containing 10 to 20 per cent of iron. These alloys, which are somewhat cheaper because of this addition, are suitable for operating at dull red heat.

From such materials an improved series of alloys has been developed, capable of withstanding high stresses at high temperatures. These alloys, known as the 'Nimonic' series, were developed during the Second World War for use in jet turbine aero-engines, providing an example of operation under the most severe conditions of high stress and high temperature. In addition the alloys must withstand corrosion by the products of combustion in the engine.

The alloys, in the following order, are suitable for operation at progressively higher temperatures and stresses. In Nimonic

75, the 80/20 binary alloy is strengthened by the precipitation of titanium carbide caused by the presence of about 0·4 per cent of titanium and 0·1 per cent of carbon. In Nimonic 80A the maintenance of mechanical properties at even higher temperatures than can be used with Nimonic 75 results from the presence of 2 per cent of titanium and 1 per cent of aluminium, the optimum properties being obtained by heat-treatment for 8 hours at 1,080°C followed by ageing at 700°C. Nimonic 90 is a nickel–chromium–cobalt alloy to which small additions of titanium and aluminium are made. A recent alloy in this series is Nimonic 105, which has outstanding resistance to creep at temperatures of 950°C; this alloy is somewhat similar to Nimonic 90 but contains molybdenum, together with a larger percentage of aluminium. Nimonic 115 is a further development of Nimonic 105.

The Nimonic alloys form a very interesting series showing what can be achieved as a result of long and intensive laboratory and engineering research for one particular application, namely the gas turbine engine. The earlier alloy, Nimonic 80A, is strong enough to withstand a force of 85 newtons per sq. mm for 1,000 hours at 815°C, whilst Nimonic 115 can withstand 240 newtons per sq. mm for 1,000 hours at 815°C, or 85 N/sq. mm at 950°C. Considerable quantities of these alloys are used in all the British gas turbine engines: Nimonic 75 for the flame tubes and the 105 or 115 Nimonics for the moving blades in every production aircraft turbine in Great Britain.

Nickel–Molybdenum Alloys

A range of nickel-rich alloys which are finding increasing use in the chemical and petroleum industries on account of their high corrosion-resistance and strength are those of nickel with up to 30 per cent molybdenum. These alloys sometimes have additions of chromium. Alloys in this range have outstanding resistance to hydrochloric, sulphuric and phosphoric acids, chlorine gas, and other acid and oxidizing solutions.

Nickel–Copper Alloys

In Canada, the copper and nickel in the ore were found to be in the proportion of two parts nickel to one part copper, together with small amounts of other elements. The metals are normally separated but, in 1905, Robert C. Stanley suggested that the mixed ore could be smelted direct, thus yielding a 'natural alloy'. This was given the name 'Monel' after Ambrose Monell, then President of the International Nickel Company. Monel contains small quantities of iron, manganese, silicon, and carbon in addition to the main elements. There was a demand for corrosion-resisting alloys and since the time of its inception Monel has been used for pump shafts and bodies, rivets and other fastenings, and components for chemical and similar plant. It is also used for the welding of cast iron. Subsequently, modifications of Monel were introduced, one, known as K-Monel, containing aluminium and titanium. The mechanical properties of this alloy can be improved by heat-treatment, giving a strength of more than twice that of mild steel. Its best-known use is for propeller shafts of powerful motor boats. This strong, corrosion-resisting alloy is non-magnetic so it also features in the construction of parts of aircraft near the compass, and in non-magnetic vessels for marine survey work and mine-sweeping, where steel, being magnetic, would influence the reading of the instrument.

A copper–nickel alloy containing 45 per cent nickel and 55 per cent copper possesses a high electrical resistance and a very small variation in resistance with change of temperature. It is used for making electrical resistance units for control switchgear and other electrical apparatus which must not be affected by temperature changes.

The cupro-nickels containing 10 to 30 per cent of nickel (the rest being copper) possess a remarkable capacity for cold work. The important use of the 70 per cent copper, 30 per cent nickel alloy for condenser tubes is referred to on page 273. An alloy containing 10 per cent of nickel and small percentages of iron and manganese is widely used for large-diameter pipelines carrying sea water. Cupro-nickel alloys, in particular 25 per cent nickel, 75 per cent copper, are required for coinage; the present British 'silver' coins are made of this alloy, as is the American 5-cent piece or 'nickel' and the Hong Kong dollar. In addition to such familiar coins, cupro-nickel is used for the Brunei 5 sen, the Gambian 50 butut, the Nigerian 10 kobo and the Malawi kwacha.

Two thousand years before nickel was isolated, an alloy called 'paktong' was made in China by smelting a mixture of what are now known to be a zinc ore and a copper–nickel ore. It was not until 1849 that a range of alloys of the same type was produced in Europe from the constituent metals nickel, copper, and zinc. Up to 1914 such alloys were generally known as 'German silver', the name then being changed to 'nickel silver'. A number of these alloys are available with nickel contents from 5 to 25 per cent, the colour ranging from nearly white with the high nickel alloys to pale yellow with those of low nickel content. The combination of attractive colour, malleability, and resistance to corrosion makes these alloys popular for tableware, having the name 'E.P.N.S.' (electroplated nickel silver) which is nickel silver, plated with a thin coating of silver. The nickel silver alloys are also used for decorative and architectural purposes. An alloy with 18 per cent of nickel is widely used for contact springs in telephone exchanges, on account of its springiness and resistance to corrosion.

There are many other interesting alloys having outstanding and often unique properties for which nickel is an essential constituent. In a list of metals given at the beginning of this book nickel is called the 'versatile metal' and this name is fully justified by its value in many diverse fields.

16
MAGNESIUM

Magnesium has been called the lightweight champion among metals. Aluminium is more than one and a half times heavier, iron and steel are four times, and copper and nickel five times heavier. This low density, combined with a relatively high strength, has won for magnesium its present place in industry.

Although a comparatively late arrival in the world of engineering metallurgy, magnesium was isolated as long ago as 1808, earlier indeed than aluminium, and efforts to produce it commercially were made throughout the nineteenth century. The reasons for the tardy development of this metal, so lavishly provided by nature (it is the fourth most abundant metal, and it is found in the sea as well as in the earth) are partly technical and partly economic. On the one hand extraction problems had to await the development of industrial electrolysis, and on the other hand there was little demand in industry for that lightness which is the metal's most characteristic property. Magnesium affords a good example of metal responding to the needs of man and fitting into the pattern of his civilization. It is significant that magnesium ores, unlike those of the older metals, are so widely distributed throughout the earth as to preclude the possibility of establishing a monopoly in them but the metal itself is so difficult of extraction as to be exploitable only by highly industrialized communities.

Of all the common metals there is none about which there is so little public knowledge as magnesium. This is because magnesium is seldom employed for articles in common use, such as kitchenware and furniture, so that we rarely have a chance to handle it. We come into frequent contact with iron, steel, aluminium, copper, zinc, lead, tin and nickel, and can usually distinguish one from the other, but if we see a piece of magnesium, we probably mistake it for aluminium. There is one important exception: in the powder form, magnesium metal is known to everybody as a photographic flashlight. Therefore the only fact about magnesium known to the general public is that it burns, and during the war this knowledge was thoroughly confirmed by experience with incendiary bombs. Consequently, the average man has difficulty in believing that so flammable a metal can be used for any structural purpose, and he is quite incredulous when told that it has been employed for cooking-pots and frying-pans. The apparent contradiction is easily explained by the fact that before magnesium can be made to burn vigorously it must be melted and in contact with ample supplies of air.

The special methods required for smelting and refining magnesium, casting the alloys, and for their later fabrication into various finished forms, were pioneered by the Germans, although a great deal of valuable development work was done in Great Britain and America. There has been a remarkable increase in output over the past seventy years. In 1900 annual world production amounted to only ten tons. This increased to 1,000 tons in 1920, to 20,000 in 1937, and to a maximum of 238,500 in 1943 – an elevenfold increase in six years, due to wartime needs of the aircraft industries. After the war production fell to about 10,000 tons per annum but after 1949 it rose rapidly and reached 215,000 tonnes in 1970. Since then the tonnage has risen to about 270,000.

Although pure magnesium has no great strength – about 110 newtons per sq. mm – suitably alloyed and worked its strength can be doubled and even trebled. Until quite recently the alloying elements most widely used have been aluminium and zinc as hardeners to raise the mechanical properties, and manganese to improve corrosion-resistance. The best type of casting alloy of this kind contains 8 to 10 per cent aluminium with 0·5 per cent zinc and about 0·25 per cent manganese. An alloy in this range of composition can be improved further by heat-treatment.

More recently it has been found that an addition of less than

1 per cent of zirconium has the property of reducing in size the individual grains of which the alloy is composed. For example, the average grain diameter of unalloyed magnesium may vary from one to fifty millimetres, while the same metal containing 0·65 per cent zirconium has an average grain diameter of less than one tenth of a millimetre. The smaller-grained metal is considerably stronger and more ductile. By the addition of other elements such as zinc, thorium, silver and rare earth metals, a number of alloys having desirable properties such as strength at room and elevated temperatures, creep resistance and good castability, have been developed. Although the potential importance of zirconium as an alloying metal for magnesium was originally discovered by German metallurgists, the difficult alloying technique was perfected in Britain, and work on the new alloys proceeded simultaneously with the development of jet propulsion.

Further research work during the Second World War showed that thorium additions improved the resistance to creep of the magnesium–zinc–zirconium series of alloys, particularly in the cast state. The room-temperature mechanical properties are also generally superior to earlier alloys. Castings in magnesium alloys containing about 0·7 per cent zirconium, 3·0 per cent thorium, and 2·5 per cent zinc are used in both British and American jet engines.

Another range of important magnesium alloys contains about 0·7 per cent zirconium, 2 to 5 per cent zinc and 2 to 3 per cent cerium. A recent invention enables very strong and ductile castings to be made by heat-treatment in hydrogen which permeates the alloy and modifies its microstructure. This is an unusual use of hydrogen since in most metals it reduces mechanical properties.

Magnesium alloys can be subjected to all the usual metallurgical treatments, including casting, diecasting, rolling, forging, extruding, and pressing. In spite of its flammability in the molten state magnesium can be cast with ease and safety if certain precautions are taken to prevent the metal from burning. Magnesium is used mainly in the cast form.

No metal can be cut, filed, drilled or shaped so easily or so fast as magnesium. This of course helps to reduce the cost of the final product. Magnesium also lends itself readily to welding and riveting. The strength of cast and wrought magnesium alloys in relation to their weight is very high indeed. The cast alloys can show strengths of 300 newtons per sq. mm and wrought alloys up to 380 newtons. The strength-to-weight ratio of such materials resembles that of a high tensile steel.

The principal engineering advantage of magnesium alloys is their lightness, and the main field of application is therefore transport, motor-cars and aircraft. During the war retractable undercarriage parts were cast in magnesium, pilots' seats were made of welded magnesium tubes, petrol tanks of magnesium sheet, while a large number of engine parts, such as super-charger blower casings, were also produced from magnesium alloys. In Britain alone, during the Second World War, nearly a million aircraft wheels were made of magnesium. It is also used for crank cases and gear boxes in cars and heavy commercial vehicles and for wheels in high performance cars. Other uses include the bodies of portable automatic tools, for many components of textile machinery, for cameras, office machines, and for some items of military equipment.

The predominant use of magnesium in the motor-car industry has been by Volkswagen, which at one time consumed 38,000 tonnes per annum, nearly one-sixth of the world's magnesium output. The famous 'Beetle', for example, contains about 22 kg of magnesium castings, the air-cooled engine design being ideally suited for ultra-light alloy. The reduction in output resulting from high fuel costs and the trend to front-mounted water-cooled engines has now reduced Volkswagen's consumption of magnesium. Its modern uses are mainly in gear boxes and other transmission components but the pressure diecasting plants operated by Volkswagen remain prime examples of highly developed technology.

It is interesting that quite apart from the large consumption of magnesium by Volkswagen, the other uses of the metal in Germany per head of population are twice that of U.S.A. This is because once magnesium is accepted as a viable material of construction its use spreads. A similar problem is evident in

Britain. In fields where magnesium is accepted, textile machinery, aircraft, chain saws for tree felling, the development of magnesium naturally grows. Other markets, the British motor trade for example, are conservative.

Potentially, magnesium could replace aluminium in many applications, the choice between the two metals depending only on commercial considerations. However the relative prices of the two materials are not the only factors. The machining of magnesium is much easier and faster than aluminium. On the other hand one has to allow for the extra cost of a chromate chemical finish, to improve corrosion resistance. Magnesium is a chemically reactive metal, though it does not corrode as much as iron and steel. To prevent deterioration, magnesium and its alloys can be protected by special methods to make them resistant to corrosion. These protective processes consist of immersing the pieces to be treated in vats containing a hot solution of chromate salts. Black or golden oxide films, containing chromic oxide, are thereby deposited on the surface of the magnesium and these form a protective layer which acts as a base for the application of paint for further protection.

Large quantities of pure magnesium are used for alloying with other metals, as a deoxidizer or refiner in melting metals such as nickel, and in powder form in fireworks. Another use of magnesium is in the treatment of cast iron to nodularize the graphite, which makes the iron stronger and more ductile (see also page 171).

On page 271 reference is made to the 'sacrificial protection' of iron by zinc. For some applications magnesium is more effective, and it is used extensively in America to protect oil pipelines and steel water tanks by connecting them to magnesium anodes. Another electrochemical property of magnesium is used for photo-engraving; indeed this method has already been used for book printing instead of the mono- or linotype methods.

A large amount of magnesium is required for alloying with aluminium, specially for the alloy with $4\frac{1}{2}$ per cent magnesium which is used for easy-open tops of tin plate cans. At present about 18,000 tonnes of magnesium are consumed for this one purpose, and it is estimated that by 1978 about 27,000 tonnes may be required. However, as we explained on page 189, this mixture of several metals makes it difficult to reclaim scrap and it may be that other inventions will supersede the tear-off tops.

At present the magnesium industry, like many others, is in a state of some indecision. Magnesium should be amply available – a cubic kilometre of sea-water contains over a million tonnes, but about 15,000 kilowatts of electricity are required per tonne of the metal, and electric power is not sufficiently plentiful these days. Recent developments in extraction technology have been centred on the use of natural brines containing magnesium chloride; these provide for production and sale of the chlorine formed as a by-product amounting to two kilograms of chlorine to every kilogram of the metal. Some of the plants had teething troubles but the fundamental problems associated with the process have been solved and by 1976 all the companies concerned were operating or constructing pilot facilities capable of manufacturing several thousand tonnes of magnesium per annum. By 1977 or 1978 some at least of these companies will be producing several tens of thousand tonnes per annum from new, large-scale plants.

17

SOME MINOR METALS

ANTIMONY. In appearance antimony is similar to zinc. Being hard and brittle (it can be powdered with a hammer), it is rarely used alone, but is employed as a useful hardening addition to lead, chiefly in printers' type metal, solder, white metal bearings, accumulator plates, telephone sheathing, lead-filled bullets and in shotgun ammunition.

ARSENIC. This element is not a true metal, though it has some metallic properties; it is classed as a semi-metal or metalloid. It is alloyed with lead for sporting ammunition shot and, in conjunction with antimony, in lead cable sheathing. Arsenic and some other elements, including indium and gallium, are used in the manufacture of transistors.

BARIUM. This silvery-white metal is spontaneously flammable in moist air and, though valuable in the form of some of its chemical compounds, has only few uses in metallurgy, for example in German lead alloy bearing-metals, in association with calcium.

BERYLLIUM. Beryllium is lighter than aluminium and nearly as light as magnesium; it has a melting point of 1,278°C – much higher than that of aluminium. It has good corrosion-resistance in air at temperatures below 500°C, also a high strength-to-weight ratio. It was considered to be a wonder metal with a promising industrial future, for the construction of lightweight structures such as aircraft fuselages and possibly for missile bodies and space vehicles.

The ores of beryllium are sparsely distributed and the methods of extraction extremely elaborate, all of which make it costly. Unfortunately, difficulties have been encountered in processing the metal to develop it into suitably workable material; consequently interest in industrial applications has developed slowly. The raw material has to be converted to a metal ingot by vacuum-melting in an electric induction furnace. Beryllium is then machined down to a fine powder, which is heated and compacted in a vacuum by hot-press sintering into the appropriate shape of rods, bars, or tubes. The handling and fabrication is made additionally complex because beryllium has toxic properties. It is a costly metal, usually bought in the form of hardener alloys which are sold at prices depending on the amount of beryllium included. For example a 5 per cent beryllium aluminium hardener alloy costs about £140 per kilogram of beryllium; a kilogram of this 5 per cent alloy would therefore cost about £7.

The most important alloys containing beryllium are those with 0·5 to 2·0 per cent of the metal and copper as the main constituent. Precipitation heat-treatment will give hardness of the order of 350 Brinell, a figure far superior to that of only about 45 for pure copper. An alloy of copper with 1·8 per cent beryllium, which has been rolled, heated to 800°C and quenched in water, has a strength as great as mild steel. If this quenched alloy is now re-heated to 335°C for two hours and then cooled, precipitation hardening takes place and the alloy develops a strength two or three times greater than that of mild steel. This 'beryllium copper' is among the strongest of non-ferrous alloys.

BISMUTH. Apart from the use of medicinal compounds of bismuth, this element is chiefly used as a constituent of fusible alloys (see page 65) and low-melting-point solders.

BORON. This is another semi-metal, possessing both metallic and non-metallic properties. Principally it is used in the form of borax (sodium borate combined with molecules of water) and boric acid, in the manufacture of porcelain, enamels, glass, in dyeing textiles, fire-proofing timber, as a food preservative, and for pharmaceutical purposes.

In ferrous metallurgy small quantities of boron improve 'hardenability' (the degree to which steel will harden when quenched) and is introduced into the molten steel as the readily soluble alloy ferroboron. Metallic boron has a great capacity to absorb neutrons, and boron steel is utilized for control purposes in nuclear power stations (page 308).

Extreme hardness is a quality associated with the metallurgical uses of boron, and relatively inexpensive steels may be surface 'boronized' to give a hard-wearing surface often superior to that produced by carburizing or nitriding. Compounds formed between boron and various metals, e.g. titanium and zirconium, have the advantage of a hardness not appreciably below that of diamond, are better conductors of electricity than many pure metals, do not melt below about 3,000°C, and are extremely resistant to corrosion by molten metals and salts. They are now becoming commercially available and are expected to find many applications where these properties are required.

CADMIUM. This is one of the few metals that is not commercially extracted for itself, but is always a by-product. It occurs in zinc ores in the proportion of about 1 part of cadmium to 200 of zinc, and is separated during the refining of that metal. The biggest use of cadmium is in the form of the yellow/red pigments, cadmium sulphide and sulpho-selenide, seen in many plastic products. The next major use is as an electroplated coating on steel components, especially those needing to be soldered. Nickel/cadmium rechargeable batteries power miner's lamps and various portable electrical appliances and large versions are used to start heavy duty diesel engines. Cadmium alloys are used as aluminium solders and for bearings. The use of cadmium to strengthen overhead copper electric cables is described on page 203.

CAESIUM. This metal possesses the highest thermal expansion – about eight times that of iron and steel and a hundred times that of Invar (page 228). Apart from mercury, caesium has the lowest melting point of all metals: 28·6°C. It is used in 'atomic clocks' and has enabled physicists to measure time intervals and frequencies to an accuracy of one part in 10,000 million.

CALCIUM. This metal is soft and corrodes rapidly in air, so is not used on its own. Recently, however, it has found a useful application when alloyed with lead, where as little as 0·04 per cent brings about an age-hardening effect; the strength of the lead is thereby nearly trebled. The importance of calcium has increased since it is used in the process of winning uranium metal from the ore. Some of calcium's chemical compounds are very well known – for example limestone and chalk are calcium carbonate, but the 'chalk' used to write on blackboards is calcium sulphate.

CHROMIUM. This metal is familiar as an electroplated coating over nickel plate, on steel, brass, or zinc alloy. The chromium layer, only about 0·002 mm thick, gives a hard bright surface which improves the durability of the article. The most important use of chromium, however, is in alloy steels, where quite small amounts confer increased hardness after heat-treatment. In amounts of up to 12 per cent, it improves the resistance to scaling or oxidation of iron at high temperatures. There are several low and medium alloy steels containing chromium with other elements, such as molybdenum and vanadium, which can be used continuously at temperatures from 500°C upwards to 700°C, for components of thermal power stations and steam turbines.

Additions of over 12 per cent chromium give the range of modern stainless steels (page 176). Chromium in conjunction with nickel is used to strengthen and harden cast iron. When required as an addition to cast iron or steel, the chromium is provided as ferro–chromium, containing 70 per cent chromium, 4 to 6 per cent carbon and the rest iron. Ferro–chromium has a lower melting point than that of the pure metal and is a cheaper and more convenient way of adding it to steels and cast irons. Chromium is also used in making nickel–chromium alloy resistance-wires for electric heating elements. One well-known alloy, composed of 80 per cent nickel and 20 per cent chromium, is mentioned on page 229. This range of alloys was the base from which the famous 'Nimonic' alloys were developed, for jet aero-engines.

In Australia research work has been proceeding for several years to improve those properties of chromium which at present restrict its workability and usefulness. The metal could be attractive for rotor blades of gas turbine engines, since it has a high melting point, a good oxidation-resistance, and a lower density than some of its potential competitors such as tungsten and molybdenum. One of the difficulties, however, is that chromium is rather brittle at room temperature. It has been learnt that the brittleness is due to a combination of factors such as: gaseous impurities, particularly nitrogen; grain boundary films formed by carbon or sulphur; and finally surface condition and contamination. No really satisfactory method has yet been found to overcome all these difficulties to give a suitable commercial alloy, and the work is still incomplete.

COBALT. The alloy 'Stellite' (cobalt, chromium and tungsten) is used for the rapid machining of hard metals, since it maintains its hardness and cutting edge even at red heat; it is also used to build wear-resisting surfaces on valves and valve seats. This series of alloys has been modified for work at high temperature, a typical alloy containing 62 per cent cobalt, 28 per cent chromium, 5·5 per cent molybdenum, 2·5 per cent nickel, and small amounts of iron and carbon. It is used for various parts of gas turbine engines. One of the most important modern uses of cobalt is as the bonding agent for tungsten carbide in tipped tools as described on page 291.

An alloy containing 57 per cent iron, 35 per cent cobalt, 2 per cent chromium, 5 per cent tungsten, and 0·9 per cent carbon was, until recent years, the most highly magnetic material that had been discovered. It has been superseded by a series of nickel-iron–aluminium alloys, whose inception has been responsible for improvements in aircraft instruments, magnetos and cycle dynamos. The most magnetic material yet discovered is a compound of cobalt with copper and the 'rare earth' metal samarium.

GALLIUM. A soft silvery-white metal, gallium is sufficiently widely distributed to encourage endeavours to find markets for it.

It is unusual in having a wide range between its melting point, about 30°C, and its boiling point, about 2,070°C. Gallium, like arsenic, is used in 'doping' silicon transistors. Compounds of gallium with arsenic and phosphorus have been used in the development of tiny electro-luminescent 'lamps' which can be used in clusters to form digital read-out displays, or individually, in conjunction with a light-sensitive cell, as counting devices or burglar alarms.

GERMANIUM. This is a semi-conductor, with a marked crystalline form. It is one of the materials used for transistors, which are amplifiers of small size and high reliability, using only a millionth of the power required by a thermionic valve. A transistor lasts indefinitely and its use has become widespread. Transistors were developed as a result of research in the late nineteen forties, in which semi-conductor materials were 'doped' with minute amounts of other elements, producing an enormous change in their electrical properties. Electrodes were attached to a single crystal of germanium, which was doped with arsenic at one electrode and with gallium or indium at another. This 'diode' has the property of different electric conductivities in opposite directions. By fixing two of these diodes together, back to back, a transistor is formed.

As will be seen when we discuss silicon, that material has replaced germanium for many transistors, but such is the ever-widening growth of the industry that germanium is still a significant transistor element.

GOLD. The most malleable and one of the most ductile of all metals, gold can be drawn to wire only 0·006 mm in diameter or beaten to leaf only 0·0001 mm in thickness – so thin that it is translucent. It has complete resistance to atmospheric attack and oxidation. Pure gold is too soft to use on its own and it is therefore hardened by the addition of alloying elements. The most common addition to gold is copper; the gold coinage alloy minted in Britain contains 91·66 per cent gold and 8·34 per cent copper, giving a sufficiently hard material with a slightly deeper colour than pure gold. In 1968 the Royal Mint struck over four million sovereigns, which have an intrinsic value greater than £1 sterling and are not issued in Britain. Sovereigns in fact appear to be exported for purchase by collectors or for use as a kind of international currency in Middle and Far Eastern countries.

The gold content of jewellery alloys is designated by the carat system. The term refers to a 24th part; thus 18 carat contains 18 parts of gold and 6 parts of alloying elements. The compositions used in the U.K. for jewellery are 22, 18, 14 and 9 carat. These are legal standards and gold offered for sale must be submitted to one of the four Assay Offices for checking the gold content and hall-marking to one or other of these standards. Hall-marking is the oldest consumer protection system in the world, having been carried out by the Worshipful Company of Goldsmiths in the City of London since A.D. 1300. Sometimes the gold content is expressed as a decimal; thus 9 carat is 0·375.

In addition to copper, the other elements used in gold jewellery alloys are silver, nickel, palladium and zinc. Copper alone produces reddish alloys and silver alone greenish to white alloys. Together they produce the familiar rich yellow colour; in the lower carat qualities the colour is improved by zinc additions. Palladium and nickel are used to produce the white gold alloys which are now much used as substitutes for the very expensive platinum. Rolled gold, which is used in jewellery and spectacle frames is a base metal: brass, bronze or nickel clad with 12 or 14 carat gold alloy, the composite being rolled or drawn to the required form.

The use of gold in dentistry is relatively small in Britain, but is popular in other countries. For example in the late 1960s about 20 per cent of Japan's annual consumption of fabricated gold was in dentistry. The alloys, somewhat similar to those used in jewellery, are cast and heat-treated. The metal is used in the form of gold plating in electronics for its resistance to corrosion and oxidation. Gold has the highest reflectivity for infra-red radiation of all metals and for this reason is used for coatings on space satellites, turbo-jet engine tail pipes and reflectors for infra-red lamps.

A common and important use of the metal is in gold pen nibs, usually of 14 carat gold, tipped with a bead of hard alloy containing one or more of the metals osmium, iridium or ruthenium.

The standard ingot, measuring 175 × 90 × 40 mm, weighs 400 troy, or 439 avoirdupois ounces, 12·45 kg. With rapidly changing market values, it is hard to 'price' a gold ingot. For some accounting, a troy ounce is still calculated at about 40 dollars, so at that rate the ingot would cost only about £11,000. However, if you were allowed to buy an ingot it would cost at least 130 dollars per troy ounce – 52,000 dollars, or more than £39,000. The U.S.A. Federal Bank has reserves equivalent to 700,000 gold ingots.

HAFNIUM. This metal occurs, in very small quantities, in zirconium ores. The two elements are chemically similar and it has been very difficult to separate the hafnium from zirconium; the need to do this is briefly explained on page 266. Having separated hafnium, it was realized that it is a very effective metal for absorbing neutrons, making it specially suitable for nuclear-reactor control rods (see page 314).

INDIUM. In 1928 the world output of this metal was only one gram. Though it is now produced in some quantity, indium still ranks as rare and precious. This metal is becoming increasingly important for the surface coatings of bearings, to give improved resistance to corrosion and wear. A three-layer bearing is produced; the back, of low-carbon steel, is lined with a copper–lead–tin alloy which, after broaching or boring, is plated with a layer of pure lead which is machined accurately to size; a thin layer of indium is then electrolytically deposited on the lead. The bearing is heated and the indium diffuses through the bearing, almost to the interface between the steel back and the lining.

The function of indium is threefold. Firstly it reduces the susceptibility of the copper–lead layer to the corrosive attack of lubricating oils; secondly it provides an increased fatigue strength to the lead overlay and thirdly it provides good anti-friction properties. This composite bearing material, copper–lead with diffused indium, will operate at bearing pressures of 600 kg per sq. cm in endurance tests of over 1,000 hours. Since 1956 the use of indium-faced bearings has extended to medium-heavy diesel engines and in the coming years bearings larger than 200 mm, the present manufacturing limit, will be achieved.

IRIDIUM. Having a specific gravity of 22·56, iridium, a member of the platinum group of metals, is second only to osmium in heaviness. Iridium's corrosion-resistance is even better than that of platinum, but despite a very high melting point of 2,454°C, it is less oxidation-resistant than either platinum or rhodium. Iridium is difficult to work and only recently has begun to be used in its elemental form. Iridium-tipped sparking plugs have been developed to replace platinum-tipped plugs in aero-engines. Platinum alloys containing up to 30 per cent iridium are used for jewellery; the 10 per cent alloy is used for diamond settings in U.S.A. Platinum–iridium alloys are used for standards of weight and length because of their great stability and permanence.

LITHIUM. This is the lightest of all metals, having only half the density of water, but it reacts so rapidly with air and water that it cannot be used as a constructional material. Lithium hydroxide is contained in special-purpose lubricating greases which retain their consistency over a wide range of temperature and which are unaffected by water. It is therefore possible to make a multi-purpose lubricating grease which is effective in arctic climates and at temperatures over 100°C, as well as in very wet conditions. Lithium carbonate is used on a small scale in pharmaceutical chemistry, and in the manufacture of glazes. During the Second World War, the compound lithium hydride was provided as a source of hydrogen for filling balloons in remote battle areas.

Metallic lithium has been used as a de-oxidizing agent in alloy manufacture, for lithium–lead alloys used in radiation shielding, and as aluminium–lithium alloys having high strength at elevated temperatures. The metal is unusual in the considerable range between its melting and boiling points. It melts at 186°C and boils at 1,336°C (compared with sodium, which melts at 97°C and boils at 880°C). This property makes molten lithium suitable as a heat transfer agent in nuclear reactors.

In nuclear fission two isotopes of lithium are used. Lithium-6 has a high ability to absorb neutrons and from it the hydrogen isotopes for thermonuclear fission are derived. On the other hand lithium-7 does not absorb neutrons in nuclear reactors.

MANGANESE. This is one of the most ubiquitous of the metals that are used almost exclusively for their beneficial alloying properties. Thus practically all steels contain small additions of manganese, of the order of 0·5 per cent. Each year about 10,000,000 tons of manganese ore are mined, containing about 35 per cent metal. Generally the metal is extracted as ferro-manganese, an alloy of iron and manganese; in this form manganese is generally added to steel. However, ferro–manganese contains carbon and silicon; these impurities would be harmful if added to some grades of low-carbon stainless steel so for that purpose pure manganese is extracted by electro-metallurgical methods.

Manganese is an important addition to some magnesium and aluminium alloys, where it increases corrosion-resistance. 'Manganese bronze' is a high-grade 60/40 brass containing manganese which toughens and strengthens the alloy. In addition to the main uses outlined above, there are many other alloys which embody manganese as an important constituent; these include alloys of aluminium, nickel, silver, titanium, alloy steels, including manganese steel (page 175), copper-base electrical resistance alloys and corrosion-resistant alloys.

MERCURY. The only metal that is liquid at ordinary temperatures is mercury or 'quicksilver'. It is used in making the chemical compound mercury fulminate ($Hg(ONC)_2$), which is one of the 'initiators' used to detonate explosives. Mercury also appears in various scientific instruments such as thermometers, barometers, discharge lamps, vacuum pumps, and in many forms of small electric contact-breakers. Metals such as tin, silver, and gold dissolve in liquid mercury to form alloys or 'amalgams' which are widely used in dentistry.

MOLYBDENUM (commonly called 'Molly'). Though molybdenum was first isolated in 1790, it has only come among the ranks of important minor metals in comparatively recent times. The metal has a very high melting point, 2,625°C, which is exceeded only by four other metals: tungsten, rhenium, tantalum, and osmium. It is used in the manufacture of electric lamps, wireless valves, and electric contact points, and for a growing number of rocket components. It is mechanically strong, light in weight, and has an electrical conductivity equal to about one-third that of copper. These properties also enable it to be used successfully for radiation-shielding in vacuum melting furnaces, which are rapidly increasing in both size and importance.

The most important use of molybdenum is as an addition to alloy steels, particularly in association with nickel and chromium. This combination is widely applied, though manganese–molybdenum steels are also becoming increasingly used. Development work in the U.S.A. has led to molybdenum replacing tungsten in high-speed steels. It is generally reckoned that it has twice the 'power' of tungsten: thus if a high-speed steel previously contained 18 per cent tungsten, substitution can be made by using a steel with 9 per cent molybdenum.

Molybdenum can also improve non-ferrous alloys, while for industrial and motor-car lubricants which have to withstand high pressures and temperatures, molybdenum disulphide is a useful addition. Several of the properties of the metal give it great potential in the growing nuclear industry, as a reactor construction material.

NIOBIUM. The largest consumption of this metal, known to the Americans as Columbium, is indirect – as ferro–columbium additions in steel-making. Like most 'new' metals niobium has been known for a considerable time but a full appreciation of its potential value became apparent only when major advances in extraction and fabrication techniques enabled metallurgists to provide the metal for its most exciting uses. Niobium has a high melting point of 2,468°C and the development of its alloys has now reached the stage where mechanical properties at 1,200°C are sufficient for components of space craft, jet engines and guided missiles.

Niobium is still an expensive metal, costing at least £25 per kg and as much as £100 per kg for the most sophisticated alloys in bar form. However as its production increases the price is likely to fall. Niobium alloys have great resistance to heat. The leading alloy contains about 17 per cent tungsten, $3\frac{1}{2}$ per cent hafnium and 0·1 per cent carbon. Special coatings are needed to provide oxidation resistance at high temperature.

Niobium does not absorb neutrons readily, thus making it suitable for fuel cans and support brackets in nuclear reactors. The resistance of pure niobium to the attack of liquid sodium-potassium alloy leads to its use for fuel canning material in liquid-metal-cooled reactors.

Some niobium alloys and compounds become superconductive at very low temperatures, typically below −263°C (10K). They lose all resistance to electric current. The materials most used are a niobium alloy with 44 per cent titanium, in the form of fine filaments in a copper matrix, or the niobium–tin intermetallic compound Nb_3Sn, as a tape with copper backing. These super-conductors are used for the construction of very large or high field electromagnets.

OSMIUM. The main interest of this rare metal lies in its neck-and-neck contest with iridium to be the heaviest of all elements. When a small porosity-free piece of osmium made by arc melting in a vacuum is tested, its specific gravity is 22·59 grams per cubic centimetre. A piece of iridium sheet has been measured with a specific gravity as high as 22·56. On this count osmium wins. However when the theoretical specific gravities of the two metals are calculated from the lattice spacings, osmium would have a specific gravity of 22·60 and iridium 22·65. Until a piece of iridium has actually been produced heavier than osmium, we, and no doubt the *Guinness Book of Records*, will award the trophy to osmium. In its elemental form osmium is virtually useless, as it has a volatile oxide which forms and volatilizes at room temperature. Alloyed with iridium, with which it sometimes occurs naturally, it forms very hard corrosion-resistant alloys which are used for pen-nib tips and instrument pivots.

PALLADIUM. A member of the platinum family of metals, palladium is the one most similar to platinum, though it is slightly less corrosion-resistant and has a lower melting point. Palladium is considerably cheaper than platinum, has a lower density and can replace it in some applications where conditions are not extreme. It is widely used for electrical contacts in speech circuits of telephone systems. Palladium catalysts are used in oxidation and hydrogenation processes, in margarine manufacture and the production of ethylene from acetylene. This metal has the unique property that it can rapidly absorb up to 900 times its own volume of hydrogen. Super-purity hydrogen for chemical and metallurgical processes is obtained by passing impure hydrogen through a diaphragm of heated palladium which acts as a filter, excluding other gases. In recent years a series of palladium-containing brazing alloys have been developed which are particularly suitable for joining stainless steel and cobalt alloys, as well as the nickel-chromium high temperature alloys such as the Nimonics.

PLATINUM. Platinum has exceptional resistance to corrosion, being superior even to gold in this respect; it is completely free of oxidation and has a high melting point of 1,772°C. It therefore finds application for parts where even the slightest corrosion or oxidation would be detrimental, for example, for electrical contacts carrying small voltages, and crucibles and other pieces of apparatus used in critical chemical analyses. For many years just a curiosity, platinum alloyed with copper, iridium or ruthenium eventually became popular as a jewellery alloy particularly for the mounting of diamonds. As industrial uses have multiplied however, short supply and high prices have very much reduced the amount of platinum used in this way.

Platinum catalysts are used in many chemical processes, the most important of which is the catalytic oxidation of ammonia to form nitric acid, which uses large woven gauzes of platinum–10 per cent rhodium alloy. The catalytic reforming of low octane petrol used platinum a few years ago, but non-precious catalysts have now partially replaced it in this field.

The efforts to reduce pollution and smog in the U.S.A. will

involve platinum catalysts to eliminate hydro-carbons and carbon monoxide from car exhausts. Ford's have planned to use this method in California in the mid-seventies and to extend its application later. General Motors, however, are also investigating base-metal catalysts which will be cheaper. If platinum anti-pollution devices are fitted on even a part of the world's automobiles, the effect on the price of platinum and its availability for other uses will be considerable. At the peak of the development of the aero piston engine, the biggest single use of platinum was for the tips of sparking plug electrodes. With the coming of the gas turbine engine, this application has declined, though the remaining piston engines are still dependent on the reliability of platinum-tipped sparking plugs.

Platinum and platinum–rhodium alloys are widely used in high-quality and high-production-rate glass-making processes. Optical glass is melted in large platinum-lined crucibles to avoid contamination by refractories; platinum-clad apparatus is used in many other glass-making processes. Continuous filament glass fibre is made by melting the glass in solid platinum–rhodium alloy vessels which are heated by electrical resistance, the glass filaments being drawn off through nipples formed in the bottom of the vessel. The most accurate methods of measuring high temperatures up to about 1,400°C are either platinum resistance pyrometers or platinum/platinum–rhodium alloy thermocouples. These are widely used for the direct measurement of the metal temperature in steel making.

Platinum was originally found as native metal in alluvial deposits principally in Colombia and the Yukon, but today the only important supplies outside of Russia are from South Africa, where it is recovered from a sulphide ore, and Canada where it is found associated with the Sudbury nickel/copper ores.

PLUTONIUM. This was first made synthetically in March 1941 by bombarding uranium atoms in the cyclotron. Such bombardment leads to the formation of neptunium, which is unstable and is then transformed into plutonium. During the operation of a nuclear reactor, plutonium is formed from uranium-238 (see page 311); it can be separated chemically from the uranium and is then available for fast reactors. Plutonium is fiendishly toxic, highly reactive and a bone-destroyer. It is more fissile than uranium-235 and is an alternative or substitute for it in atomic bombs. Because plutonium is a fissile material, it has a 'critical mass' (see page 301) at which the fission chain reaction becomes self-sustaining and the hazards of radiation and explosion become imminent. The critical mass depends on the geometry of the system; it is higher for a flat shape than for a sphere. It also depends on the nature of the material, whether the plutonium is in the form of metal, oxide or solution such as plutonium nitrate. The critical mass for plutonium is smaller when it is in solution than when it is in solid form.

POTASSIUM. This is a soft silvery-white metal, with chemical properties similar to those of sodium. It has little importance in industrial metallurgy.

RADIUM. This extremely rare, highly radioactive, heavy metal is extracted from uranium ores, where it occurs in the proportion of about one part radium to 3,000,000 parts uranium. The fantastic scarcity of radium may be assessed from the production of Canada, the world's largest source of supply, which amounts to a few kilograms per annum, accurate production figures being almost impossible to ascertain. The metal is chiefly used for medical purposes on account of the penetrating radiation that it emits continuously, which is used for the treatment of cancer and some skin diseases.

Very minute amounts of radium compounds are mixed with zinc sulphide; the radiation causes the sulphide to become luminous and this is used for dials of watches, compasses, and also in luminous gun sights. This accounts for about a tenth of the output of radium. The historical interest, the medical value of radium, and the inspiration of the story of Marie Curie's endeavours are very great; but metallurgically radium has little significance.

'RARE EARTH' METALS. From time to time mention is made in technical journals of elements with tongue-twisting names, like gadolinium, which are arousing new interest. They represent about one-sixth of all the known elements but are scarce, expensive, and produced in only small quantities. A list of their names is as follows:

cerium	gadolinium	neodymium	terbium
dysprosium	holmium	praseodymium	thulium
erbium	lanthanum	promethium	ytterbium
europium	lutetium	samarium	yttrium

Scandinavian scientists in particular have played an important part in isolating these metals and the names given to several of them reflect this influence. The mineral gadolinite, named from the Finnish chemist Johann Gadolin (hence the name gadolinium) was discovered near the Swedish town Ytterby and from that the names of the four metals ytterbium, erbium, terbium and yttrium were derived. Holmium was derived from Stockholm, thulium from Thule, the early name for far-northern Europe.

The Russian mine officer Samarski lent his name to posterity with the metal samarium. Lutetium was named from Lutetia, the ancient name for Paris, and europium got its name from Europe. Cerium was named after the asteroid Ceres.

Some of these metals borrowed Greek roots for their names. Dysprosium means 'the one hard to get at'; lanthanum 'the hidden one'. The name didymium, meaning 'twin', was given to a material which was found to be composed of two elements which, when isolated, were called neodymium, the new twin, and praseodymium, the leek-green twin.

While they do not hold an important commercial position in the metal world, the rare earth metals may be expected to mark up small but notable achievements in the future. Potential fields for development include small additions to some alloy steels, components in nuclear energy plant, solar energy devices, and electronics.

Most people will sometimes have held a rare earth metal in their hands since cerium is the main constituent of cigarette lighter flints, in the form of 'misch-metall', an alloy of cerium and other rare earth elements. When this alloy is rubbed by the steel wheel of the lighter, a substantial spark is formed; the alloy is said to be 'pyrophoric'. The oxide of cerium is employed in glass-polishing and lens-making.

The family of rare earth metals has been termed a Pandora's box, but it has been pointed out that so far the list does not include either 'delirium' or 'pandemonium'.

RHENIUM. This is a very heavy, silvery-white metal with the second highest melting point, 3,167°C, and considerable resistance to corrosion. So far it has very limited industrial uses, though it has featured in marine-engine magneto contacts.

RHODIUM. This is another metal of the platinum family and is the most corrosion-resistant of them all. It is considerably lighter than platinum, but with a rather higher melting point. It is, however, somewhat susceptible to oxidation at very high temperatures and is more difficult to work than its sisters platinum and palladium. The principal use for rhodium is as a hardening alloy addition to platinum alloys, some of which are used to make the spinnerettes required in the manufacture of glass fibre and synthetic textile materials.

Rhodium is the easiest of the platinum metals to electro-deposit and with its white colour, relatively high reflectivity and extreme corrosion resistance, is widely used in this form, particularly on white gold jewellery. Rhodium plating is used on silver jewellery, but does not find favour as a tarnish-protective coating on larger items of silverware and flatware as the appearance, though of good white colour, is inferior to silver because the reflectivity is reduced. In the industrial field rhodium plating is used for electrical contacts and plugs and sockets, due to its great hardness allied to its corrosion resistance.

RUTHENIUM. This is the least known of the 'platinum group' of metals. Like osmium and iridium, ruthenium is difficult to work and there is at present little use for it. With its relatively low density and price, however, it is an effective substitute for iridium as a hardening addition for platinum and palladium jewellery alloys. Ruthenium electro-deposition processes have developed in recent years and although the dark colour precludes the use of these deposits for decorative purposes, their hardness and corrosion resistance approaches that of rhodium.

SELENIUM. When light falls on selenium, its electrical conductivity increases by many orders of magnitude, depending on the intensity and wave length of the light. Selenium is not unique in this respect; all semi-conductors exhibit the effect to some extent. Advantage is taken of this property in the operation of camera 'magic eyes' and Xerox copy machines. Temperature measurement devices which base their action on registering the amount of light emitted by a hot material, rely on selenium.

SILICON. Although not a metal, silicon was bound to be mentioned many times in this book because of its important role in ferrous and non-ferrous metallurgy. It is a semi-conductor element with chemical properties similar to those of carbon. As silicon occurs widely, in combination with other elements, particularly oxygen, in sands, clays, and other minerals, it is not surprising that when metals such as iron, aluminium, and copper are extracted from their ores, silicon often accompanies the metal as impurity. When iron is made in the blast furnace, the pig iron contains as much as 3 per cent silicon. Although much of the silicon is removed when pig iron is converted into steel, it confers improved elasticity; steel containing over one per cent silicon is used, in conjunction with manganese, for car springs and bridges. The aluminium alloys containing silicon have been referred to on page 186.

With the development of chemical techniques and the procedure of zone-refining (defined on page 338), silicon of exceptionally high purity can be produced for transistors; it is also used for solar cells. The development of the silicon transistor has been extraordinary. A colour T.V. set contains over two hundred silicon transistors; a computer several million. They are included in the controls of cars, refineries, trains and spacecraft.

Important decreases in the size and cost of these controls have been made possible by the use of wafers of silicon cut from a single crystal. In a similar way to the treatment of germanium, transistors are formed on chips of the silicon wafer by doping it with arsenic and indium. A tiny complete 'integrated' electronic circuit, containing ten thousand or more transistors and resistors, can be grown on one such monolithic chip.

SILVER. Although this metal is resistant to many corrosive agents, it is attacked by sulphurous fumes which is the reason why silver articles tarnish and blacken in industrial atmospheres. Like gold, silver used for jewellery, teapots, forks and the wine labels one hangs on decanters, is subject to hall-marking. In the United Kingdom there are two legal silver alloys, 'Britannia silver' containing 95·84 per cent silver and the much more common 'Sterling silver' containing 92·5 per cent silver. There is no restriction on the other alloying elements, but these invariably are copper with or without cadmium.

The 'silver' coinage of Britain nowadays is a copper–nickel alloy containing no silver, but prior to 1947 silver alloys were used as shown in the following table.

Table 20. Composition of British Silver Coinage

Years	Silver per cent	Copper per cent	Nickel per cent	Zinc per cent
1921	92·5	7·5	—	—
1922–6	50·0	50·0	(various silver alloys were used in this period)	
1927–47	50·0	40·0	5·0	5·0

A historic use of silver coinage remains in Britain. Maundy money, distributed by H.M. the Queen to the poor on the Thursday before Easter, contains 92·5 per cent silver. These coins are now in new pence 1p, 2p, 3p and 4p. The amount given is determined by the Monarch's age, one penny for each year, and the number of recipients is governed by the same factor, one woman and one man for each year.

Industrial uses of silver include the lining of vats and other components used in dairy, brewing, vinegar and other food industries due to the metal's high resistance to attack by acetic acid and other organic corrodants. Electro-deposition consumes large quantities of silver, both for decorative purposes and in applications such as lighting reflectors and wave guides, making use of the metal's properties of the highest reflectivity to light and highest electrical conductivity of all metals. The electrical and thermal conductivity of silver leads to its wide use in electrical contacts, particularly where high currents and voltages are involved. Silver solder and brazing alloys are discussed on page 280.

Every photograph depends on the light sensitivity of silver bromide and, as the photographic industry is expanding, the requirements of silver for that use alone is considerable. On the other hand silver coinage is declining in favour of cupro–nickel. The table below shows the uses of silver for 1970, and the estimate for 1980. The figures are for all countries except those behind the Iron Curtain, which do not provide statistics for us.

Table 21. Silver Consumption in Troy Ounces
(one million troy ounces equals about 30,000 kilograms)

Use	1970 *(in millions)*	Estimated for 1980 *(in millions)*
Industrial	280	520
Coinage	55	25
Photography	140	250
Total	475	795

SODIUM. This is a very soft, wax-like metal, which reacts vigorously with water and corrodes rapidly in air. Liquid sodium has proved valuable for extracting the heat which is evolved in the fast nuclear reactor of which the prototype was built at Dounreay in the north of Scotland. Sodium melts at only 90°C and it is a superb conductor of heat from the reactor core, transferring that heat to the steam generators. An important use of sodium in the aluminium–silicon alloys was described on page 186.

TANTALUM. Though this element had been separated in an impure form early in the nineteenth century, it was not till 1903 that the German scientist Werner von Bolton isolated pure tantalum. From its initial limited use in metallic lamp-filaments, tantalum has grown into an important modern metal. Its three outstanding characteristics, high melting point, excellent corrosion-resistance, and ease of working, have assisted this growth. In its natural environment, tantalum is always found in association with niobium; a series of skilful chemical processes is needed to separate these two metals effectively. Tantalum metal, besides having the very high melting point of 2,996°C, is unaffected by the majority of acids and consequently is used to avoid many corrosion problems in the chemical, pharmaceutical, and other industries. It is used in surgery for bone-splints, screws, brain clips, and other pieces which must be left permanently in the body. Because of its corrosion-resistance and its appreciably lower cost than platinum, tantalum is being employed in chemical equipment where heat has to be transferred under intensely corrosive conditions.

There has been a recent development of tantalum, in powder or foil form, for making small capacitors in the electronics industry. By using tantalum, it is possible to get capacitors of one-tenth the size of the normal aluminium-foil ones, and still maintain the same capacity. Other applications include metallic rectifiers and similar semi-conductor devices. The compound of tantalum and carbon is added to improve the performance of tungsten and titanium carbide cutting-tools. Tantalum carbide is the most refractory substance, having a melting point over 4,000°C.

TELLURIUM. When less than one-tenth of one per cent tellurium is alloyed with lead it increases the hardness, strength, and corrosion-resistance. Tellurium is sometimes added to copper alloys and, in conjunction with lead, to steels to promote free machining properties. Tellurium is not a true metal, but is more correctly described as a semi-metal. It is mainly produced as a by-product.

THORIUM. This soft and ductile metal is closely related to uranium and is one of the very few naturally occurring elements which could possibly be utilized to develop atomic power by fission of one of its isotopes. Apart from possible developments as nuclear fuel its main use is for adding to magnesium to improve the metal's properties at high temperatures.

TITANIUM. As long ago as 1791 titanium was recognized by William Gregor, an English clergyman and mineralogist. Lately it has ceased to be a laboratory exhibit and its industrial development has been spectacular; three tons were produced in 1948 and about 25,000 tons in 1957. That was a peak year; in 1964 about 11,000 tons of titanium sponge were produced (this is the condition in which the titanium is obtained from the ore) and from this 9,000 tons of wrought titanium were made. In 1970 the figures for the free world production were 23,000 tons of sponge and 18,000 tons of wrought titanium.

The technical improvements in titanium came as a result of a concerted effort by chemists and metallurgists, particularly in the U.S.A., to extract the metal from the ore in a pure enough form to provide alloys which could be fabricated and which possessed the properties required by aeronautical engineers: lightness, corrosion-resistance and strength. Titanium is widely distributed in nature; of the truly structural metals only aluminium, iron, and magnesium are more abundant. The processes of extraction, melting and fabrication are all expensive because of the need to avoid contamination by oxygen, nitrogen, and carbon – all elements which are present in the air. The metal or its alloys cost from about £3·50 per kg for forging billet, up to about £16·00 per kg for alloy sheet.

The extreme reactivity of titanium in the molten state makes it impossible to melt in normal crucibles because no refractory will resist it. Consequently a radically different melting technique had to be devised. The type of furnace known as a consumable-arc furnace is shown in Plate 17 and uses a direct-current arc between an electrode made of the metal to be melted and a starting slug resting on the base of a water-cooled copper crucible. The entire arrangement is enclosed in a vessel which either contains a non-reactive gas such as argon or which may be completely evacuated. The electrode is usually made of compressed blocks of titanium sponge in the form in which it is purified; alloying additions are included at this stage. Normally a slender ingot is melted first to act as the electrode for a second melting operation.

This development work has led to the perfection of techniques and equipment which can help both new metals and old ones. Thus consumable-arc melting furnaces are now commonly used for zirconium, molybdenum, hafnium, chromium, high grade nickel alloys for jet engines and for alloy steels, where great reliability is essential. In spite of their reactivity, titanium and its alloys are being cast directly to finished shapes for chemical, engineering and aerospace uses.

The alloys of titanium are classified according to the structural phases that are present. This subject was discussed on page 83, in connection with the structure of brass. Titanium alloys fall into three groups; the single-phase alpha alloys, two-phase alpha–beta, and single-phase beta alloys. The first group includes 'pure' titanium (containing oxygen, nitrogen and carbon); titanium alloyed with 5 per cent aluminium and 2·5 per cent tin; and titanium with 2 per cent of copper. The second group is the largest; a typical alloy contains 6 per cent aluminium and 4 per cent vanadium. Another popular alloy contains 4 per cent aluminium, 4 per cent molybdenum, 2 per cent tin and 0·5 per cent silicon. The third group of titanium alloys is less developed than the first two, although the ability to cold form these 'beta alloys', followed by heat treatment to high strength, has led to the development of several alloy compositions for aircraft rivets and sheet.

Current 'commercially pure' titanium has a high ductility so that it can be readily shaped; its tensile strength ranges from 300 to 750 newtons per sq. mm. Some titanium alloys can be heat-treated to over 1,400 newtons per sq. mm and are used for forged parts where maximum strength per unit weight is required, for example in vertical take-off aircraft. Although titanium alloys have relatively good fatigue properties these are partially diminished in practice because of the somewhat higher sensitivity, compared with alloy steels, to the presence of notches causing stress concentrations. At one time it was hoped that, because of the high melting point of titanium, 1,670°C, it would be found to have high creep-strength at high temperatures. The best so far achieved would be able to stand service at 550°C only.

Aircraft speeds have increased very rapidly, owing to the considerably greater thrust possible from gas turbine engines; this means that skin temperatures of aircraft fuselages have become much higher than in previous designs. American and British designers foresee titanium being a major metal for aircraft frames and skins in the future. The Anglo-French Concorde, flying at Mach 2·0, uses only about 4 per cent of titanium, mainly in the engine surrounds; the Lockheed A-11 interceptor fighter, capable of over Mach 3, manufactured in the mid 1960s is said to contain up to 85 per cent structural weight in titanium alloys. Current designs of aircraft have between 7 and 26 per cent structural weight in titanium alloys, the high figures being typical of military applications. Plate 18 shows the RB211 turbo-fan, which uses titanium for compressor discs and blades, shown on the left of the picture.

Titanium is establishing many uses for highly stressed components of missiles, rockets, and space capsules; other typical applications are for blading, discs, rings and ducting in the compressor stage of many jet engines. It is being increasingly used for air frame structural forgings such as flap tracks, brackets, engine mountings and undercarriage components, as well as for engine cowlings, fire walls and leading edges. In the U.S.A. it is also used for airborne military equipment, such as mortar bases and armour-plating.

The outstanding corrosion-resistance of titanium has caused growing interest in its use in bone surgery where its lightness combined with strength, availability in wrought forms, and complete resistance to body fluids gives it advantages over alternative materials. Titanium is resistant to a wider range of corrosive substances than austenitic stainless steels, particularly in seawater; furthermore it forms anodic films. Use is being made of titanium in chemical plant, particularly bleaching plant for the textile industry, reaction vessels, parts of chlorine cells, for racks, hooks, and heating coils in electroplating vats and for oil refinery pipe work.

Very recently tool-tips coated with a thin layer of titanium carbide have become established throughout the metal-cutting industry. At the same time solid titanium carbide tool-tips have been developed in the U.S.A. Both these new products are competing strongly with tungsten carbide.

An important use of titanium is as a grain refiner for alloys in aluminium, nickel, and iron, particularly during welding operations; but for none of these applications does the metal need to be extracted in its elemental form. By far the largest use of titanium is in a non-metallic form, as titanium oxide for paints, where it has a greater whitening effect than lithopone or than white lead carbonate because of its permanence and its ability to reflect white light.

TUNGSTEN.* The drive to higher and higher operational temperatures for metals, particularly in rockets and jet engines, has focused the spotlight on tungsten, which has the highest melting point of any known metal (about 3,400°C). Unfortunately it has some disadvantages including its extremely high density; tungsten is two and a half times as 'heavy' as iron. Its resistance to oxidation is not outstanding at high temperatures. Furthermore tungsten has a most volatile price structure; the sharply fluctuating demand for the main outlets of the metal – cutting tools and mining equipment – causes large variations in the price of tungsten. Because of its high melting point, tungsten is often used in the electrodes of welding equipment. The filaments in electric light bulbs are made of tungsten.

The metal was incorporated in the original 'self-hardening steels'. The property of alloys steels of suitable composition to self-harden was first discovered by Robert Mushet in about 1868, and led to the manufacture of 'Mushet's steel' for cutting

* At a meeting of the International Union of Chemistry, held at Amsterdam in 1949, it was decided that this metal should be renamed 'Wolfram'. Since then a further meeting decided to revert to 'Tungsten', but the chemical symbol is W.

tools. It was some years later, in 1900, before a really satisfactory steel for machining at high speeds was produced, containing 14 to 18 per cent tungsten. Steels of similar composition are used for hot-working metals for components of extrusion, forging, or diecasting dies. Many of the present-day high-speed and hot-die steels contain lower amounts of tungsten, 4 to 6 per cent, but a few per cent of molybdenum, chromium, and vanadium may be added.

URANIUM. The uranium ores most familiar to mineralogists are pitchblende and carnotite, which contain uranium oxide, U_3O_8. In 1974 free world reserves of U_3O_8 mineable at under £9 per kg were estimated at 950,000 tonnes; in that year about 25,000 tonnes of the oxide were produced. The U.S.A. has large deposits in Utah, Wyoming and Colorado; Canadian uranium ore comes mainly from the Elliot Lake area; South Africa produces uranium as a by-product of gold mining and a major new mine is being developed in South West Africa. In the next few years France will add to her own supplies those from her ex-colonial territories in the Gabon, Niger and the Central African Republic. Recent discoveries could make Australia a very large producing country by the end of the decade. Deposits of mineable uranium ore usually contain up to 5 kilograms U_3O_8 per tonne but a discovery at Nabarlek in Queensland is reported to contain 23·5 kilograms per tonne.

The Czechoslovakian St Joachimstal mines, recently renamed Jachymov, are famous because they provided Marie Curie with a supply of residue for her three years' task of i solating radium. The demand for the metal now is mainly derived from its use in the production of nuclear energy for peaceful purposes. Demand for uranium was about 16,000 tonnes in 1970 and may rise to 60,000 tonnes by 1980. Increase in demand for uranium, has led to sharply rising prices throughout 1974, from £8 per kg to £14 per kg. The use of uranium for the production of nuclear energy is examined in chapter 21 and some of the metallurgically irritating problems of this metal are also mentioned.

VANADIUM. This element is used principally as an alloying addition to iron and steel, acting as a scavenging agent to remove non-metallic impurities, but it has been found to be beneficial in certain alloys in other respects. A variety of tool, die, and cutting steels contain 0·1 to 0·5 per cent vanadium in addition to larger amounts of chromium (1 to 2 per cent), about 0·6 per cent manganese, and sometimes molybdenum and tungsten. By appropriate heat-treatment, fine-grained steel of exceptional toughness is obtained.

Steels containing 0·2 per cent vanadium and about 2 per cent chromium with 0·45 per cent carbon are widely used for the dies from which zinc alloy pressure diecastings are made, though chromium-molybdenus-vanadium steels, heat-treated are more suitable for aluminium diecasting. Demands for stronger aircraft structures have led to the production of 'super steels' (see page 180) having strengths of the order of 1,500–2,200 newtons per sq. mm in strip and sheet form. One such alloy contains 0·5 to 1·0 per cent vanadium with 5 per cent chromium, 1·3 per cent molybdenum, 1 per cent silicon, 0·3 per cent manganese, and 0·4 per cent carbon.

Vanadium is also added to cast irons because it produces a uniform hardness in die blocks, gives a maximum depth of hardness in chilled iron rolls, and excellent high-temperature behaviour for pistons for diesel engines. For all these uses the amount of vanadium present is of the order of 0·1 to 0·5 per cent. A titanium alloy containing 6 per cent of vanadium and 4 per cent of aluminium is used, in the U.S.A., in the form of forgings and sheet, for aircraft and space missile components.

ZIRCONIUM. Although this metal was isolated over 150 years ago it remained a laboratory curiosity till recently. In the form of finely divided powder or swarf, zirconium ignites at low temperatures and finds application in photoflash bulbs, where it produces a very white light, making it suitable for colour photography. Because of its corrosion-resistance to alkalis, the metal is used in the rayon industry for spinnerettes. An alloy of copper with about 15 per cent zirconium has very great strength; and the importance of zirconium as an alloying element in magnesium alloys is mentioned on page 235. Zirconium is valuable in nuclear engineering because of its low neutron absorption. The main outlet for zirconium is the alloy with 1·5 per cent tin, 0·1 per cent iron, 0·15 per cent chromium and 0·05 per cent nickel, used as a fuel canning material in water-cooled nuclear reactors, including those in submarines.

Till recent years, zirconium as produced from its ores was contaminated with small amounts of the rare metal, hafnium. When the possibilities of zirconium in nuclear engineering were appreciated, it was realized that a hafnium-free metal must be produced, otherwise the impurity would cause slowing down of neutrons. This has now been achieved and the hafnium-free zirconium is being used increasingly.

18

CORROSION

Corrosion is the destructive attack upon a metal by agents such as rain, polluted air, or seawater. The rusting of iron and steel provides the best-known example of corrosion, and the continuous painting of steel bridges and ships illustrates that protection against rust is an ever-present problem. A Government Committee recently conducted a survey into corrosion and protection in the United Kingdom. In its report, published in February 1971, the cost of corrosion in the U.K. was estimated to be £1,365,000,000 per annum, excluding costs in the farming area. The annual expenditure on protective coatings of all kinds was estimated at £772,000,000. The annual bill for the whole world is thought to be at least £30,000 million.

Many metal articles are used under conditions where they are affected by the atmosphere and by moisture; hence corrosion by the attack of these two media is the most familiar kind. Sometimes the conditions in which a metal gives service are of a severely corrosive nature; for instance, the steel-work of a pier at the seaside tends to become corroded, since the intermittent immersion in seawater with the rise and fall of the tide causes even more intense corrosion than would occur if the steel were covered permanently by the sea. The speed of corrosion is affected also by marine growths such as seaweed, barnacles, even by bacteria. Extreme cases of rapid corrosion are often caused by sewer water or by the effluvia and contaminated waters produced by factories. A railway line in a rural district may last seventy years but in tunnels the life may be only three to eight years. Yet not all corrosion is reckoned as undesirable; for example, the appearance of bronze statues or copper roofs is often enhanced by a patina of a bluish-green corrosion-product, known as verdigris.

The problem of corrosion may be approached by first considering the chemical action whereby the surface of a metal is attacked by gases in the atmosphere, which form chemical compounds with the metal. For example, one might imagine atmospheric oxygen gradually converting the surface of a piece of iron into iron oxide. However, atmospheric attack cannot be so simple as that, for if iron is stored for long periods in contact with dry air it does not rust so much as would be expected from the usual behaviour of the metal. If the iron is kept moist, rusting proceeds rapidly, which might lead one to expect that water was the influence causing the corrosion. Yet if iron is kept in air-free distilled water, rusting does not take place so fast as when air is dissolved in the water. Even the combined presence of air and moisture does not fully account for the formation of rust, for pure iron is attacked less than commercial grades of iron or steel. Such observations as these lead to the discovery that corrosion is stimulated when a metal is non-homogeneous and when it is in contact with water containing some gaseous, liquid, or solid substance.

Various theories have been put forward to account for the vagaries of corrosion, but facts such as those given above support the idea that a relationship exists between corrosion and electrolysis. In other words, a corroding metal behaves as though it is a kind of electric cell or wet battery. At first the relationship between the rusting of a nail and the action of a battery may not seem obvious, but it has been proved that corrosion is usually accompanied by the setting-up of small localized electric currents. In a battery an electric current is produced by suspending two metals in a chemical solution. When the circuit is completed one metal, known as the anode, dissolves, while an electric current flows through the solution from this corroding metal to the other, called the cathode. One may imagine a piece of metal in contact with moisture behaving like a small battery; the presence of particles of an impurity, or contact with some other metal, allows a difference of voltage to be set up, thus causing a minute electric current to flow. The moisture, containing air or some dissolved chemical substance, conducts the electricity and local attack is begun.

The great difficulty in studying corrosion is that so many variable factors are likely to influence the initiation, course, and final result of corrosion. For example, the purity of the metal, the composition and inter-relation of all the substances with which it comes in contact, the presence of bacteria, the possibility of small externally produced electric currents being present, may all affect the corrosion.

Sometimes the corrosion behaviour inside a crack or recess in a metal article provides an example of what is known as 'differential aeration'. Moisture which has penetrated into the bottom of the crack becomes devoid of dissolved atmospheric oxygen, while at the mouth of the crack the water contains plenty of dissolved oxygen. This provides an electrical potential difference, that is an anode at one end and a cathode at the other. Under such circumstances, the lower areas devoid of oxygen become the anode and they corrode.

The factors are so difficult to control and classify that it is not surprising that a vast amount of exploratory work has had to be done on the subject. In the early days of this field of research British workers were very much to the fore. Among others may be mentioned Dr Ulick R. Evans of Cambridge University; the late Dr William H. J. Vernon, of the National Research Laboratory,* Teddington; Dr John C. Hudson, of the former Iron and Steel Institute Corrosion Committee; and the late Dr Guy D. Bengough. In the United States one of the foremost workers is Professor Herbert H. Uhlig of the Massachusetts Institute of Technology.

METHODS OF COMBATING CORROSION

The more that can be learnt about the causes and mechanism of corrosion the easier it will be to prevent its damaging action; several main possibilities offer themselves towards this end.

Pure metals are likely to resist corrosion better than metals containing impurities. Thus pure aluminium resists attack better

* Now part of the National Physical Laboratory.

than a commercial variety; 'chemical lead' of high purity is more resistant to corrosion than a less pure grade. The work being done on the extraction of high-purity metals is likely to help in combating the wastage due to corrosion. The elimination of impurities from alloys usually helps to improve their corrosion-resistance; in a given series of alloys, those which contain only one structural constituent are likely to resist attack better than those which contain more than one constituent.

The search for alloys with superior resistance to corrosion is a never-ending attempt to satisfy an insatiable demand. In ordinary structural steels quite small additions of copper of the order of 0·05 per cent improve the resistance to corrosion. Such improvements, though significant, are nothing like so powerful as the effect of additions of 18 per cent chromium and 8 per cent nickel, which makes steel stainless under most normal atmospheric and other corroding conditions.

If the initial coating of corrosion product is adherent and impervious it can prevent further attack from taking place. Aluminium provides an illustration of this, for it forms a thin but strong film of aluminium oxide on its surface which restricts further corrosion. Stainless steel containing chromium is similar, its corrosion-resistance being due to the formation of a protective oxide film. On the other hand, ordinary iron and steel are liable to progressive rusting because the layer of rust on the surface is porous and tends to flake off, so further rusting penetrates inwards, though often at a reduced rate.

The metal may be coated, to prevent the access of corrosive agents. For example, protective coatings of other metals can be applied by electroplating, dipping, spraying, or by cladding one metal with another, as when sheets of aluminium alloy are clad with pure aluminium. Most iron and steel, however, is protected by paint, bitumen, or proprietary preparations. The use of synthetic resins in paints has increased enormously, thus giving a long life to the paint film; finishes consisting of plastic materials and rubberized paints have also been introduced. Each type of finish of protective coating has its own particular aesthetic or economic merit.

Engineers often have the opportunity to prevent metal from

being used in unduly corrosive conditions. For example, it would be folly to build a power station on a site where it would draw cooling water from a river into which another factory is discharging corrosive solutions, although such things have been done in the past.

CATHODIC PROTECTION

Since corrosion is associated with the setting up of small electric currents, it is sometimes possible to reduce corrosion by creating conditions under which an electric current in the opposite direction is formed in the corroding metal. This method, called 'cathodic protection', is illustrated by the galvanizing or zinc-coating of mild steel. The zinc corrodes in preference to the steel, the positive current flowing from the steel to the zinc. The zinc coating slowly disappears but it has delayed the rusting of the steel. This effect is aptly described as 'sacrificial protection'. A similar system operates for the protection of underground pipe lines. These are usually wrapped with impregnated cloth or given a bitumen coating but it is often necessary to provide additional protection at gaps in the wrappings or at pores in the bitumen. This may be achieved by burying magnesium anodes in the soil close to the steel pipe and connecting them to it by an insulated wire. The steel becomes cathodic to the magnesium and current flows along the wire to the magnesium, which corrodes instead of the steel and thus has to be replaced periodically.

A third form of cathodic protection is applied, particularly to steel structures in seawater; a direct electric current, closely controlled in potential and amperage, is passed through the seawater to the steel structure. The protective current has to be conveyed through an insoluble anode, usually of graphite, lead, iron–silicon alloy, or through a titanium rod or plate thinly coated with platinum. Many of the piers and jetties erected by oil companies for mooring and unloading oil tankers are protected in this way; for example at Thameshaven. It is interesting however that piers at seaside resorts are not so protected. They were built many years ago and ample metal was included in the supporting columns to allow for loss of metal by seawater corrosion.

In contrast with such methods of cathodic protection, it is possible to build up an oxide surface layer by making the metal the anode in a special electrolyte, when a controlled current is passed. The oxide layer so formed is tough, and resistant to wear and corrosion because the underlying metal is insulated both electrically and physically. Apart from aluminium (page 183), other metals can be anodically treated, including titanium and stainless steel. It is possible to get a range of different colours on these metals by varying the thickness of the oxide film, thus increasing the aesthetic appeal and enabling metal panels to be produced as coloured pictures.

CONDENSERITIS

'Our ships, great and small, have been at sea more continually than was ever done or dreamed of in any previous war since the introduction of steam. Their steaming capacity and the trustworthiness of their machinery are marvellous to me, because the last time I was here one always expected a regular stream of lame ducks from the fleets to the dockyards, with what is called "condenseritis", . . . but now they seem to steam on almost for ever.' These words were spoken by the then Mr Winston Churchill in presenting the Navy Estimates to the House of Commons on 27 February 1940, but probably few appreciated that his remarks paid tribute to years of painstaking research work by a handful of British firms. In all forms of steamships, a 'condenser' is vital to secure maximum efficiency; it consists of nests of tubes about 4 to 7 metres long and 18 mm in diameter, with walls about 1·5 mm thick, all arranged like a gigantic honeycomb. Cold seawater is circulated through the condenser tubes while, outside the tubes, exhaust steam from the engines condenses to pure water and is returned to the boiler for generating more steam. If the tubes corrode, the seawater leaks into the boiler water, causing rapid corrosion of the whole power unit.

During the First World War, the average life of such condenser tubes, which were then usually made of 70/30 brass, was about three to six months in naval vessels. In the 1940s, the average life of the improved condenser tubes was about seven years, and that under conditions more arduous than ever before. At present the average life of condenser tubes varies from ten to twenty years according to the alloys used and service conditions. This improvement has arisen as a result of the efforts of metallurgists in industry and in research laboratories. The most far-reaching advance was made by developing a copper alloy containing 28 per cent nickel and one per cent each of iron and manganese; this cupro-nickel is used in Naval vessels and luxury liners. Independently of this achievement, it was found that the addition of 2 per cent aluminium and 0·05 per cent arsenic to brass containing 76 per cent copper and 22 per cent zinc produces a cheaper alloy than cupro-nickel but one which has a long service life. This 'aluminium brass' is used almost exclusively for merchant ships, and also in many land power stations where water conditions tend to be severe.

STRESS CORROSION

Failures due to stress-corrosion cracking or corrosion fatigue take place at stresses well below the normal tensile limit of the metal involved. As the names imply, in the first case fracture results from the combined influence of corrosion with tensile stress, while in the second case rapidly reversing stresses under corrosive conditions lead to failure.

For a given metal system there is usually a specific corrodent or series of corrodents capable of initiating cracking. Thus stress corrosion in brass takes place in the presence of ammonia, and for stainless steel the presence of chlorides constitutes the hazard. For mild steel, contact with strong caustic soda can result in 'caustic cracking' which sometimes affects the rivets of high-pressure steam boilers. The caustic agent comes from alkaline boiler waters, which can seep into and concentrate in the crevice under the rivet head. There seems little doubt that the initiation of cracking results from a process of corrosion on a microscopic scale.

Stress-corrosion failures fall into two categories: those taking place in the presence of stresses introduced into the metal during a deformation process and remaining 'locked up', and those stresses imposed by external forces applied during assembly of equipment or during its operation. Locked-up stresses can be removed by a low-temperature annealing process which results in more even distribution of the stress within the metal without lessening its strength.

An example of stress corrosion resulting from locked-up stresses is the season-cracking of brass. Many years ago it was noticed that cartridge cases stored in India cracked spontaneously during the monsoon season; the name 'season-cracking' was applied to this metallurgical epidemic, which led to some bad accidents when season-cracked cartridges 'blew back' when fired. This seemed to be due to some form of corrosion, though the actual attack on the metal was only slight. Other examples of season-cracking were noticed, and it was remarked that they occurred only when the metal had been cold-worked in its fabrication and later subjected to corrosive conditions. Eventually season-cracking was traced to the presence of unevenly distributed stresses remaining in the metal as a result of the deformation to which the metal was subjected in shaping (in this connection reference might be made to the description of the making of a brass cartridge case, on page 205). The article might remain for months or years in this state until the influence of some corrosive condition caused the pent-up stress to be released, and the metal cracked. One may not be able to prevent the articles from being exposed to the special conditions causing season-cracking, but it is possible to remove the stresses in brass by annealing at a temperature of 200° to 300°C. By this treatment the stress is relieved, so that the brass will not tend to season-crack, though the annealing temperature is not sufficiently high to reduce the mechanical strength of the metal.

CORROSION TESTS

The behaviour of metals and alloys in service can be forecast by subjecting them to accelerated corrosion tests. As a result of years of experience, these are fairly accurate, provided that control factors are taken into account. Such corrosion tests are usually conducted in closed chambers with water sprays containing sodium chloride, hydrogen peroxide, or acids, whichever causes the appropriate type of corrosion attack. In other tests sprays may alternately wet the specimen and dry it with warm air.

For other types of corrosion tests, liquids are made to flow rapidly through tubes at a constant speed, the specimens being either portions of tube or flat specimens which are strung together, and electrically insulated from each other. The specimens are weighed before and after test; they are examined for the type of attack, whether general thinning or localized and, if the latter, whether pitted or intergranular. In other tests, jets of liquids impinge on the samples of various metals and alloys being examined. As a result of many such tests, it has been found that the flow of a liquid can be quite critical, according to whether it is streamlined or turbulent. In the latter case, a phenomenon known as cavitation corrosion can develop, whereby small vacuum cavities are formed which on collapsing have the property of peeling the protective films away from the surface of the metal, thus exposing it to more corrosion.

Corrosion tests, thus, help to predict when corrosion is likely to occur, but they should be regarded as only one of the weapons needed in the everyday fight against corrosion. It is rather striking that the growth of knowledge about this very complex subject has caused an important change of attitude. At first corrosion was regarded as a necessary evil and the best that could be done was to build structures with more than ample weight of metal and apply paint liberally and regularly. Recently metallurgists, corrosion specialists, designers and engineers have gone into the attack, with the object of preventing corrosion from taking place. The fight against corrosion starts on the drawing board.

19

JOINING METALS

When a metal component is being intelligently designed many factors are taken into account: the cost of the finished article, its strength, reliability in service, and appearance. Often the facilities and skills of the manufacturer cause one or other process to be selected. Sometimes a metal component is designed in one piece and processes such as diecasting or investment casting are used. In contrast to these methods it is often found that components can be made effectively by assembling cheaply produced shapes of simple design, pressings for example. The development and appreciation of Value Analysis has caused some very refreshing 'new looks' to be given to product design. The attention which has been paid in recent years to joining processes, such as riveting, soldering, and welding, has made 'built-up' construction both rapid and efficient, and it is often cheaper than 'one-piece' manufacture.

It is a good thing that healthy rivalry exists between the various ways of joining metals. Certainly there has been spectacular progress in the joining and assembling of metal articles – progress which has been apparent in many industries, from the making of biscuit tins to the mass-production of ships. In this chapter the main types of joining processes are grouped into mechanical methods; soldering and brazing; and finally welding.

MECHANICAL METHODS

The process of riveting is pictured in *Fig. 51* though different forms of rivets, solid and tubular, are used and rivets are driven either hot or cold, depending on the metal. A number of well-known engineering structures contain large numbers of rivets;

for example, seven million were used for the Forth Bridge. In small confined spaces, as in some parts of aircraft structures, the problem of inaccessibility for riveting has been overcome by the use of tubular rivets and the invention of ingenious gadgets for carrying out the heading operation; for example, one type of hollow rivet carries a tiny explosive charge in its head; the rivet is put into position and the head heated, detonating the explosive, which causes the rivet head to bulge and thus secures it. For light assembly, pop-riveting may be employed. A stem with a head, like a large pin, is pulled quickly through a tubular rivet, which expands to seal the joint.

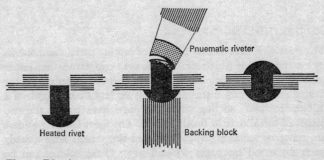

Pnuematic riveter

Heated rivet   Backing block

Fig. 51. Riveting

In addition to riveting there are other mechanical methods, including screwing and bolting which are too familiar to require detailed description, though even for these methods of assembly many improvements have been introduced in the past few years. For example, in the assembly of aircraft and motor-cars, nuts are being tightened by spanners incorporating a measuring device which shows the force being applied, thus enabling the tightness to be controlled. Many screw fasteners in use today cut their own thread on the mating part, or engage with a 'nut' having spring steel claws which close on the screw as the tension is increased. These fasteners are widely employed in locations inaccessible to a spanner. Interference fits are used to join shafts and hubs in the motor trade. Often the outer member is heated and the shaft refrigerated in order to increase the tightness of lock when the two parts revert to normal temperature.

High bond-strength adhesives have become available for joining metals. They now replace the more traditional fastening techniques, such as bolting, riveting, soldering and welding, in many fields of engineering. These adhesives are strong and versatile, and they enable metals to be bonded with many materials, including other metals, glass, ceramics, wood, concrete and plastics. Because the whole bonded area is covered with adhesive, there is no bi-metallic corrosion. The use of structural adhesives often reduces production costs compared with other joining methods.

SOLDERING AND BRAZING

The operation of soldering consists of uniting metal components by a metal or alloy which melts at a lower temperature than either piece of metal to be joined. When liquefied the solder covers the heated surfaces and forms an alloy layer less than a tenth of a millimetre thick, so that when the parts have cooled the two pieces of metal remain firmly united. The most familiar solders are those containing mainly lead and tin, which have already been discussed on pages 63 and 213. These solders can be melted with a blow lamp or heated soldering iron (the 'iron' actually being of copper).

In order to produce a satisfactory joint it is essential that the solder should wet and flow over the surfaces to be joined. The term 'wetting' can be illustrated by a simple analogy. A few drops of water placed on a clean dish will spread out and wet its surface, but if placed on a greasy dish, the water will remain as droplets. A little soap or detergent added to the water will dissolve the greasy film and enable wetting to take place. Metals which are to be soldered should be cleaned before commencing work; and, in order to maintain the clean surface and to remove oxide films, a flux of resin or zinc chloride is usually applied which further cleans the metal surface and seals it from the tarnishing effect of the atmosphere.

The ease with which a solder will wet and flow over the surface of another metal depends on the characteristics of the metal concerned. For example, molten lead will not wet a copper surface, however clean it is, but if a little tin is added to the lead the resultant alloy will readily flow over copper. Aluminium used to be regarded as a difficult metal to solder because of the tenacious film of aluminium oxide which forms on the surface and interferes with wetting. Similar problems arise with many other alloys containing aluminium; as little as 0·5 per cent aluminium makes soldering somewhat difficult. Special fluxes have been developed for soldering aluminium and, although some care is necessary, satisfactory joints can be made. Considerable use has been made of this development in the electrical industry, for making cable joints in aluminium wire cables.

For many years bicycle frames have been made by joining steel tubes to the brackets by brazing operations, the techniques becoming more sophisticated and faster over the years. Dip brazing, in which parts are immersed into molten brass, has virtually disappeared from the modern cycle works; most assemblies are now joined by internal charges of brazing brass being pre-loaded into the assembly and applying external heat.

The brazing solder is a form of brass (hence the word 'brazing') which usually consists of 60 per cent zinc with 40 per cent copper. The alloy can be melted at about 850°C by a gas torch. Such higher-melting-point solders are often called 'hard' solders, in contradistinction to the lead–tin or 'soft' solders, which melt at 200° to 250°C. A brazed joint is considerably stronger than a soft-soldered joint and, partly for this reason, there has been a tendency towards brazing instead of soldering, a trend which received further impetus owing to the high cost of tin.

One interesting technique which has arisen from the development of brazing methods has been capillary jointing, a method widely used in the manufacture of motor-car accessories, refrigerator parts, and radio sets. If two steel parts are to be joined as shown in *Fig. 52*, they are placed together with a thin disc of copper foil between them (*a*). The parts are then passed through an electrically heated furnace filled with an atmosphere free from oxygen but containing hydrogen. The copper foil melts, but is kept clean and free from oxide film by the presence of the hydrogen and so the molten copper flows in the gap; capillary attraction, better known as surface tension, pulls the molten metal up the sides (*b*). This process lends itself to mass-production methods; the join is clean, neat, and economically made.

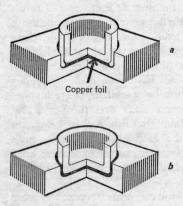

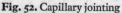

Fig. 52. Capillary jointing

Silver solders are alloys of copper, zinc, and silver, sometimes with the addition of cadmium; one silver solder contains 60 per cent silver with 30 per cent copper and 10 per cent zinc, the melting point of this alloy being 735°C. Until recently, silver solders were mainly for jewellery and other fine work, but now they are becoming used in engineering on account of their strength, fairly high melting point, and the accuracy and neatness of the joint obtainable.

Although there is an unfortunate absence of a suitable solder having a melting point between about 300° and 600°C, the solders, apart from this gap, offer a useful range of properties and melting points, from the low-melting-point lead–tin solders, to the high-melting-point brazing solders.

WELDING

Welding processes can be classified into two groups, namely those involving pressure (sometimes accompanied by melting) and those involving only fusion. The former group includes both forge- (or hammer-) welding and the various resistance-welding methods, while the latter is concerned with methods utilizing heat sources comprising oxy-fuel gases and electric arcs. The blacksmith practises a form of welding by heating two metal parts and hammering them together while they are still hot. Some metals will weld, even without heat being applied; for instance, two clean flat pieces of lead, silver, gold, or platinum can be laid face to face and hammered cold until they become welded. Each metal or alloy has a temperature above which hammer-welding becomes possible, though oxidation of the surface of the metal often interferes with the success of welding at this minimum temperature. For example, hammer-welding of aluminium is difficult, because the layer of oxide prevents a metal-to-metal contact; other metals, as, for example, cast iron, are practically impossible to hammer-weld because they are brittle.

Each link in an 'iron' chain is made from mild or alloy steel rod which is threaded through the preceding link, turned over and the ends hammer-welded together, usually automatically these days. The strength of each individual junction must be equal to that of each remaining link, and the reliability of such welds is demonstrated by the use of iron chains in ships' anchors and in lifting tackle.

WELDING BY ELECTRICAL METHODS

When an electric current is passed through two pieces of material in close end-to-end contact, the metals heat up, depending on their electrical resistance; heat is also developed at the interface due to the high contact-resistance. This fact is applied in a process known as resistance butt-welding, which is a development of hammer-welding, described earlier. The two metal parts are placed in separate current-carrying clamps and butted together while an electric current passes across the joint. Heat is developed locally at the point of contact and when the welding temperature is reached, pressure is increased, resulting in an increase in area at the interface and a joint is formed between the two components.

In 'flash butt-welding', parts to be joined are clamped in electrodes, one of which is movable. The welding current is switched on and the two parts are brought together to initiate a flashing action. This is continued until the interfaces are clean and molten at their surfaces. The components are then butted together fairly rapidly to squeeze out all the molten metal and thus produce a joint which has a characteristic forged structure and which contains no cast material.

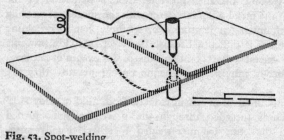

Fig. 53. Spot-welding

Another method, applying a similar principle, results in the formation of a series of small welded spots along the joint. This method is known as 'spot-welding' and is particularly suitable for the joining of sheets of metal. The two sheets are held together with two electrodes lightly pressing on either side (*Fig. 53*) and an electric current is passed through the electrodes, thus heating the metal between them; immediately afterwards the two electrodes are pressed tightly together. At the point of contact the metal parts may or may not be liquefied momentarily, but whatever happens they are joined effectively.

A pair of metal sheets can be spot-welded at as many points as are required, giving an effect analogous to a row of rivets. As the metal is continuous through each joint, the strength of a series of spot-welds compares well with that of rivets and on thin sheet has the advantage of being lighter, quicker, and easier to produce, in addition to which a more streamlined surface is obtained. A further development of spot-welding is 'seam-welding', in which a pair of metal sheets is passed between electrodes in the form of rollers so that a continuous line of weld is obtained. *Fig. 54* illustrates seam-welding, though in some machines the lower electrode does not revolve and may be replaced with a flat bar or plate.

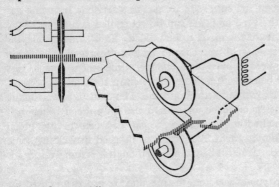

Fig. 54. Seam-welding

These types of electrical pressure-welding are particularly suited to mass-production methods. Parts of motor-cars and aeroplanes are joined by spot- or seam-welding; for example, a car may include about three thousand spot-welded joints and some aeroplanes many tens of thousands.

In recent years many new joining processes have been developed and brief reference must be made to ultrasonic welding, which appears to offer a great deal of promise for joining thin materials. In this method two metals, not necessarily of similar composition, are placed in close contact as in spot-welding and subjected to high-frequency vibrations for a short time. Special equipment is required to convert the electric energy supplied into mechanical vibrations but it is claimed such joints can be made without special cleaning treatment and with very little deformation.

FUSION WELDING

For larger joints another type of welding is used in which molten metal is run in between the joint and subsequently solidifies so as to unite the two pieces of metal. There are two chief methods of obtaining the necessary temperatures, firstly high-temperature gas flames, usually oxy-acetylene, and secondly the electric arc. By burning acetylene with a supply of oxygen the oxy-acetylene flame is produced with a temperature of about 3,000°C. By adjustment of the supply of oxygen or acetylene from the gas cylinders the type of flame can be controlled according to the requirements of the metal to be welded. Thus some metals, for example brass, weld better if the amount of oxygen is in excess of that required to burn the acetylene, while others, such as cast iron, weld better with a flame where extra acetylene is present, this being known as a reducing flame. The flame is played on the parts of the metal to be joined and also on a metal 'filler-rod' which is melted and runs into the gap between the two metal parts, so joining them. A flux has to be used in welding some metals, in particular the light alloys, though it is unnecessary for steel.

Another type of welding, the metallic arc process, is used for the welding of ships' structures, large machines, and structural steelwork. The heat is developed by striking an electric arc between the job and a metallic filler-rod. This is usually of a low-carbon steel, but additions of metallic powders and de-oxidizers through the flux coating modify somewhat the composition of the weld metal. The formation of the arc develops so much heat that the end of the rod melts and forms a weld. The rod is coated with suitable flux which melts when the arc is struck, so that the weld solidifies and cools under the protective coating of molten flux.

The technique of welding has been extended to shipbuilding and offers a rapid and effective method of construction. By the use of welding instead of riveting weight is saved, since riveted joints require overlapping of the ship's plates and each rivet adds its little load to the dead-weight of the vessel; furthermore a welded ship is harder to sink than a riveted one. The Germans were early in the field of welded pocket battleships, for, in order to make the best of restrictions imposed by various naval treaties, they used welded construction so as to reduce unnecessary weight and make it possible to mount heavier guns and obtain speeds greater than would otherwise have been practicable in a battleship of the prescribed tonnage. Large and small naval vessels have been made with extensive use of arc welding; three-quarters of the jointing work in the *Ark Royal* was welded. Perhaps this famous ship should have been called the '*Arc Royal*'! Ten million rivets were required in the *Queen Mary*, built in the 1930s, but thirty years later the *Queen Elizabeth 2* was a welded structure. The only rivets were those used to connect the aluminium superstructure to the hull.

Welding is also widely used in the construction of armoured fighting vehicles, and not only does this method facilitate production, but it helps to ensure the safety of the crew. One of the disadvantages of rivets or bolts in tanks is that the heads may be blown off when the vehicle is hit by a shell, even though it does not penetrate the armour-plate; such flying pieces of metal have been known to kill members of the tank crew.

The applications of metallic arc welding have recently been widened by the provision of a variety of mild steel electrodes to suit special purposes and welding positions. Some of these have now been developed so that excellent welds can be made with a minimum of skill.

Stainless and heat-resisting steel electrodes are marketed which are suitable for all the commercial steels and the welding of tanks for the food, brewing, and chemical industries. Electrodes are available for welding tin bronzes, aluminium bronze, nickel, 'Monel', pure aluminium, and aluminium–silicon alloys, as well as a variety of electrodes giving deposits of very great hardness.

There is increasing interest in automatic welding processes where repetition work of a suitable kind is being carried out. One such type is the submerged-arc process in which a bare wire is fed through a heap of fused glassy flux under which the arc is formed with currents up to 2,000 or 3,000 ampères. This process is useful in the welding of very thick steel plates.

One of the most interesting arc welding processes was developed in the U.S.A. and the U.S.S.R. In this, the electro-slag process, used for welding thick plate, the joint is set up vertically and the open sides are enclosed locally with water-cooled copper plates which traverse with the welding head. The electrode, in the form of wire or strip, is fed in under a flux cover to form a pool of metal, while the welding head and slides are gradually traversed upwards. The process somewhat resembles the technique used for continuous casting of metals but in this instance a consumable electrode is employed to produce the molten metal.

The introduction of inert-gas-shielded metal-arc-welding processes, for example the argon-arc process, have revolutionized the jointing of non-ferrous metals and have made it compara-tively easy to weld many metals, including aluminium, which used to be regarded as difficult to join. In addition, these pro-cesses have made it possible to weld titanium, zirconium, and tantalum, thus helping the widespread use of these metals in the chemical engineering, aircraft, and other industries.

In the tungsten inert gas (T.I.G.) process an arc is struck between the component to be welded and a tungsten electrode, held in a suitable holder, around which flows a stream of inert gas, which is usually argon. The argon shield protects the heated tungsten and the weld metal from oxidation, thus enabling troublesome fluxes to be dispensed with (Plate 20).

When welding any alloy containing substantial amounts of aluminium, an alternating current arc must be used. The arc exerts a cleaning action on the surface of the aluminium which removes the thin tenacious film of aluminium oxide and enables the edges of the materials to melt and flow together. With this method there is virtually no limit to the thickness of material that can be welded and in addition the arc exhibits characteristics which make it possible to use the process for vertical welding and for overhead welding. The vast strides that have been taken in applying aluminium to the building of ships' superstructures would not have been possible without inert-gas-shielded metal-arc welding.

The joining of dissimilar metals has always been difficult

where fusion processes are employed; this disadvantage may be surmounted by using ultrasonic vibrations, based on sound or compression waves generated at frequencies well above the audible range. By using a type of spot-welding electrode, vibrated by a 'transducer', it is possible to weld metals without fusion occurring and without any significant squeezing pressure being necessary.

Another technique is to prepare the two flat surfaces of metal and heat them under high pressure. This results in a perfect join, known as a 'pressure weld'. High-frequency induction furnaces for melting metals have been employed for some time, but recent developments point to the increasing use of high-frequency electricity as the source of heat for the rapid soldering and brazing of metals and also the hardening of steel and other heat-treatment operations. A development which may help welding in the future is the 'plasma jet'. In this type of torch high-energy inputs are possible, up to 50 kilowatts, and alloys can be fed in as powder. By this means, materials such as borides, carbides, or nitrides, which have very high melting points, can be melted and sprayed on to a backing metal such as steel.

When a variety of different shapes of metal are welded together to make a finished engineering structure it is usually called a weldment. By this means quite complicated shapes, from box girders to submarines, can be made economically from simple forms of wrought metal such as strip, sheet, plate, rod and tube.

The various methods of welding call for great skill. The operator's hands and eyes must coordinate rapidly and accurately as the weld is progressively formed along the line of the joint. Too long a pause of the flame or the arc over the joint and a large hole suddenly appears! Yet undue haste results in insufficient melting of the parent metal. And this is only the beginning of a welder's difficulties. A weld has been described as the whole of metallurgy in miniature, for it calls for a knowledge of the melting, casting, and forging of metals. A weld carried out by an inexperienced worker can be weak, brittle, coarse-grained, and can contain blow-holes. In addition to all these troubles the welder can quite easily ruin the properties of the metals he is trying to join. On the other hand, a skilled operator using suitable equipment and materials produces welds which are at least as strong as the parent metal; and the weld can be proved to be efficient and reliable by X-ray examination or by ultrasonic waves.

20

POWDER METALLURGY

Although interest in the metal platinum dates from the middle of the eighteenth century, it was known only in the form of small grains, in which condition it was discovered in nature. Attempts to melt these failed because the furnaces then available could not provide a high enough temperature. At the end of the eighteenth century William H. Wollaston discovered that larger, malleable pieces of platinum could be made by pressing the powdered metal into blocks, heating them in a coke furnace, and striking them while hot with a heavy hammer. Later a similar process was applied to tungsten, which could be produced in powder form, but which even today cannot be melted on a commercial scale.

From such beginnings powder metallurgy developed and in the last few years has enormously expanded, particularly as the result of initiative by the American automobile industry. By powder metallurgy, a new way has been provided of making either compact blocks of metal for subsequent mechanical working, or finished metal articles having specific properties which could not be produced by any other means. It may be wondered what advantages are to be derived from changing metals into powders if they are usually intended only to be compressed together again. There are several reasons.

(1) Some metals such as tungsten cannot be melted under commercial conditions, though they can be made as powders by chemical processes. Powder metallurgy enables them to be formed into solid pieces.

(2) Powder metallurgy offers an economical way of shaping small parts which can be moulded in powders in a similar way to that adopted in moulding thermo-setting plastics. This is a

building-up process, where all the metal is used and there is no machining scrap.

(3) For special purposes, non-metallic substances can be incorporated into metals. For example, graphite, an excellent lubricant, can be put into bronze which is to be used for bearings; carbon into copper for making current-collecting brushes in electric motors; and diamond powder into various metals for grinding very hard materials, such as hardened cutting tools.

(4) The techniques of powder metallurgy are leading metal-lurgists to a new way of making sheet metal direct from powder. The cost of modern rolling mills is so enormous, and so much power is required to reduce an ingot of metal stage by stage, that development work is being encouraged to produce strip direct from powder. This holds out great potential for hard metals such as stainless steel.

Several processes can be used for making metal powders. For example, the oxide of the metal (e.g. tungsten) can be heated in hydrogen; the metal (e.g. copper) can be deposited by electrolysis as a fine powder; or the metal (e.g. lead) can be atomized by pouring a molten stream into a blast of air. Having obtained a suitable powder, it is put into a die, which is shaped to the finished form of the article and the powder is then compressed at up to 800 newtons per sq. mm pressure. The die is opened and the pressed powder is found to have been consolidated, though it is not at all strong and must be handled gently. The article is then heated in a furnace, the temperature of which is about two-thirds the melting point of the metal. The heating process, known as sintering, causes the grains of metal to bind together. An alternative process consists of compressing the powder at the elevated temperature, thus combining the pressing and sintering into one operation. This hot-pressing has the advantage that lower pressures are required than for cold pressing, though the life of the die is shorter. When the tem-perature required for hot-pressing is too high for metal dies to be used, they can be made of carbon or graphite.

The strength of the common metals when pressed and sin-tered can be increased, and greater accuracy obtained in the sintered piece, by coining. In this operation, the piece is forced into a shaped die which is slightly smaller than the piece itself. A new development is impregnation or infiltration. A pressed piece of iron is sintered in such a way that it is extremely porous, and then it is placed into molten copper until all the pores are filled. Strengths up to 600 newtons per sq. mm can be obtained by this technique, which is claimed to be the most economical way of producing porosity-free parts by powder metallurgy.

For years metallurgists and engineers sought tools which would be superior to high-speed steels for cutting and machining metals. It was found that when tungsten powder was heated with carbon to a temperature of about 1,500°C a compound, tungsten carbide, was thus formed; this is exceedingly hard, and its use for cutting tools was suggested. Such tools, now used in millions, are made by mixing tungsten carbide powder with up to 25 per cent cobalt; the mixture is pressed into blocks and then heated in hydrogen to a temperature of 700° to 1,000°C so that they become about as hard as chalk. These blocks are cut and ground to the required shape and sintered by being again heated in hydrogen, but to a temperature of 1,400°C to 1,500°C, to become much harder than the hardest steel. The pieces, used for tool tips, are usually brazed on the end of steel shanks. The great hardness of sintered tungsten carbide enables drills to be made of this material, capable, for example, of drilling holes in glass. Titanium carbide and tantalum carbide can be added when the tools are required for machining steel. Tungsten carbide is used for making dies or die inserts for such processes as extrusion and wire-drawing. Powder metallurgy methods are used for die inserts in the 'compacting tools' to produce other powder metal components when these are required in big quantity.

So-called 'oil-less' or 'self-lubricating' bearings are made by sintering a mixture of copper and tin powders to which varying percentages of graphite are added. They are then soaked in oil, and no oil holes or grooves are required, since, when in use, oil will soak through such a bearing as though it were a sponge. In some cases, the amount of oil held in the bearing is enough to last the life of the machine, and some are so porous that they could be used as wicks in oil lamps. Self-lubricating bearings are used in the automobile industry, and in many domestic articles such as washing machines, vacuum cleaners, and electric clocks.

Although it is easy to melt and cast iron, it is sometimes found technically and commercially advantageous to make small and intricate iron components by powder metallurgy. An interesting application is for gears used for oil pumps in motor-cars; a gear made from iron powder by pressing and sintering accurately to size is better than one which is machined from a bar of steel, because by leaving the gear porous it soaks up oil and runs almost noiselessly. Other applications of sintered iron are small parts for adding machines and typewriters. Sintered brass is being used for parts of small locks, and an increasing use of powder metallurgy is for small magnets of the 'Alnico' type. Strong permanent magnets are now produced commercially from ultra-fine superpure iron powder particles.

There is now an important use of powder metallurgy in producing porous metal filters for handling chemical fluids, fuels, and oils; bronze, nickel, or stainless steel are used for these purposes. Particularly in the U.S.A., clutch facings and brake linings are made with iron and copper powders, together with silica and graphite. The powdered friction material is usually brazed or otherwise fixed to a steel backing during the sintering operation. High-purity nickel alloys, particularly those with high magnetic permeability, are now being produced commercially by compacting, sintering, rolling, and annealing. The absence of melting and casting in the production of such items has resulted in far less impurity pick-up than was encountered when conventional methods were employed.

Many of the so-called 'new metals', including tantalum, molybdenum, and niobium, as well as some of the older metals, such as nickel, copper, cobalt, and iron, can be extracted from their ores to yield the metal in the form of powder. The cost of such powders is not higher than solid metal and may in fact be lower. Hence there is an incentive to fabricate by powder-metallurgical methods; and 'ingots' of molybdenum and of nickel, weighing a ton or more, have been made by pressing and sintering.

Powder metallurgy is a comparatively new industry and, after a period when it was finding its feet, it is now moving ahead with considerable verve. As the processes become automated, sintered products compete with other processes. Simple components can be made at speeds of 1,500 per hour, and close limits of accuracy of 0·025 mm can be achieved. Motor-cars have an increasing use of sinterings. One model of the Xerox copier contained 89 sintered components, but this was replaced by a new model containing as many as 341 sintered parts.

Recently developed copiers have contained fewer sintered products, but those which have been used are of much greater size and complexity than before.

21

METALS AND
NUCLEAR ENERGY

In 1896, Henri Bécquerel, a French physicist, noticed that a piece of uranium mineral in his desk drawer had caused the darkening of an adjacent packet of photographic plates. This led him to realize that the uranium compound gave off radiation capable of penetrating the cardboard box and the wrapping paper. In the following year, Marie Curie, looking for a research subject for her doctorate, decided to investigate the radiation discovered by Bécquerel, which she called 'radioactivity'. Some three years later, Marie Curie and her husband, Pierre, succeeded in discovering and isolating radium which was contained in uranium ore in almost imperceptible traces, but which displays radioactivity to a far greater extent than uranium, gradually becoming transmuted into lead and helium.

From such discoveries came ever-widening knowledge, causing scientists to re-shape their ideas about the structure of atoms, unsettling previous beliefs in the permanence of matter and giving mankind the opportunity to use radioactivity for healing, to create the atomic bomb for large-scale destruction, and to offer a new source of power to augment or supersede previous supplies of energy, some of which are decreasing rapidly. The development of nuclear power has also presented metallurgists with many new and difficult problems.

Mathematicians, chemists, physicists, engineers, and metallurgists of many countries have contributed to the build-up of knowledge which made it possible to harness the immense forces available when matter is converted into energy, with which it is inter-related, as shown in Einstein's equation:

$$E = MC^2$$

where E is a measure of the energy released when a mass M is annihilated; C is the velocity of light. If the whole matter of one kilogram of a substance could be destroyed and changed into energy, over twenty thousand million kilowatt hours would be produced.

The first nuclear-powered ship, the American submarine *Nautilus*, travelled 60,000 miles and consumed only four kilograms of uranium; had it been propelled by conventional means, about 10,000 tons of oil would have been required. A modern nuclear power station of the fast breeder-reactor type will produce as much power from a ton of uranium as a conventional power station produces from a million tons of coal.

ATOMIC STRUCTURE

Atoms are very numerous and very small; in a piece of metal the size of a pin's head there are many more atoms than the total of all the letters in every book Penguin Books Ltd has ever published.

Before the First World War, Ernest Rutherford in Britain and Neils Bohr in Denmark suggested that an atom is formed from a central nucleus which is positively charged; negatively charged electrons rotate around the nucleus, but are prevented from uniting with it by its speed in orbit, just as the earth is prevented from falling into the sun. The atom of hydrogen has one electron circling around a nucleus of one positive charge, the proton. Helium has two electrons round a nucleus of two protons; lithium three, iron twenty-six and uranium ninety-two. The number of protons in the atom of an element is known as its atomic number.

The actual weights of the atoms and their constituents were not determined till the twentieth century; what are known as atomic weights in chemistry are relative numbers. It was found that electrons are only about one eighteen-hundredth the mass of protons. The electrons move around their nuclei in elliptical orbits, like planets. The orbits are grouped together in rings, and the number of electrons in each outer ring is responsible for most of the properties of the element in question. Hydrogen and helium have only one ring; lithium's three electrons arrange in two rings; aluminium has three rings; several of the well-known metals including iron, nickel, copper and zinc have their electrons dispersed in four rings; some of the heavy metals such as silver and molybdenum have five rings; gold and the rare earth metals have six rings. Uranium and radium have seven rings – the most ever observed.

ISOTOPES

When the atomic weights of the then known elements had been measured it was surprising that they were not always whole numbers; for example chlorine was found to have an atomic weight of $35\frac{1}{2}$. It had at first been hoped that the atomic weights would all be exact integers, each atom perhaps being made of an assembly of so many hydrogen atoms. The explanation of this irregularity came, in 1913, from a British scientist, Frederick Soddy, who confirmed the existence of 'isotopes'. Most chemical elements have two or more isotopes, which differ in atomic weight but which have identical chemical properties and identical physical properties, except those determined by the mass of the atoms.

In 1919 Frederick Aston was able to prove that about three-quarters of chlorine has an atomic weight of 35, while the other quarter has atomic weight of 37. This mixture gives the actual atomic weight $35\frac{1}{2}$ mentioned above. Since the chemical properties of the two types of chlorine are identical, they cannot be separated by any chemical process but Aston was able to isolate them in minute amounts by the mass spectrograph. Under an appropriate electromagnetic field, the two isotopes can be made to follow slightly different paths, because of their differing masses; they can then be separated and identified, much in the same way that two rockets of differing weight but fired off with a similar thrust will follow differing trajectories.

Although most elements are a mixture of two or more isotopes and sometimes (tin for example) as many as ten, one isotope is generally present in a preponderant amount. Thus oxygen contains

99·80 per cent of oxygen 16
0·03 per cent of oxygen 17
0·17 per cent of oxygen 18

Even hydrogen was proved to contain 0·02 per cent of an isotope with atomic weight 2, which became known as 'heavy hydrogen' or 'deuterium'. 'Heavy water', formed by uniting heavy hydrogen with oxygen, has the same chemical properties as ordinary water, though it is more expensive than that well-known liquid. It has many important uses both in atomic research and in the materials and processes used in nuclear energy.

Having proved the existence of isotopes it was illuminating that their individual atomic weights were approximately whole numbers. It should be noted, however, that their atomic weights, though very close to whole integers, are not precisely so. But why did the numbers of electrons (represented by the atomic numbers) increase in steps, one by one, but the atomic weights (represented by the weight of the nucleus of the atoms) often increase by steps of greater than one at a time? The answer was found in the suggestion of Sir James Chadwick, that the nucleus of atoms is complex in structure. Each element has a certain number of protons, the positively charged particles in the atomic nucleus, and the same number of circulating electrons. The negative electric charge on an electron is equal in magnitude to the positive electric charge on a proton. Uncharged particles, called neutrons, and each having very nearly the same weight as a proton, are associated with the proton in the nucleus. Thus, in the element carbon, there are six electrons, circulating and six protons with six neutrons in the nucleus, together mainly responsible for carbon's atomic weight of 12.

In a given element all the isotopes have the same number of electrons, which provide its characteristic chemical properties. They all have that same number of protons in the nucleus, but each isotope has a different number of neutrons. The weight difference between each isotope lies in the differing number of neutrons.

SPLITTING THE ATOM

Rutherford and his co-workers realized that under bombardment release of energy would take place if an atomic nucleus was split up, but there was a twofold problem in achieving this. First, the target was a very small one; if a single atom could be magnified to the size of a football pitch, the relative size of the nucleus would be that of a 10p coin. Secondly, if a positively charged proton were the selected projectile and if it were successful in approaching its minute target, it would have difficulty in hitting it, because positive charges repel each other. The problem might be compared with that of a blindfold man standing at the rear of a football grandstand and throwing a dart at the coin while a howling gale is blowing. So Rutherford's experiment on splitting atoms could yield only isolated, chance results. As Rutherford himself remarked: 'Anyone who looks for a source of power in the transformation of the atoms is talking moonshine'.

To achieve a bull's eye on the nuclear target, the most suitable projectiles then available were what are known as 'alpha particles', which are spontaneously thrown off from radioactive elements such as radium. They are the nuclei of helium atoms which have lost their circulating electrons and so have a nett positive charge.

By directing a beam of alpha particles, from a radioactive element, on to thin layers of various elements, Rutherford was able to witness an occasional lucky hit, which produced scintillations on a fluorescent screen. This proved that the nucleus of atoms can be partly destroyed but this was achieved in inconceivably small amounts, atom by atom. Furthermore once an isolated atom had been hit, nothing further happened to increase the chances of further hits.

Chadwick's discovery of neutrons provided a method of bombarding the nucleus of atoms without the disadvantage of the positively charged protons being repelled. As a neutron is an uncharged particle, it would not be deflected by the target. Referring to the previous illustration of the man aiming a dart at a coin on the football pitch, the use of neutrons would at least avoid the gale and thus would simplify the task of the marksman, but his target would still remain depressingly elusive.

What was needed, but what did not exist, was a method by which when one atom was split, the disintegration itself gave off further neutrons, thus causing what later was to become known as a 'chain reaction'. If, for example, two neutrons could be given off, they might lead to two other atoms being split, which would yield four neutrons, then leading to 8, 16, 32, 64, 128, and so on in rapidly increasing amounts.

Free neutrons do not exist on earth, though occasional stray ones are produced by the action of cosmic rays on matter. A beam of neutrons can be ejected from the nucleus of an element by subjecting that element to intensive bombardment by artificially accelerated charged particles. However, this process could not lead to an economic source of power because it expends much more energy than could be obtained by the impact of the neutron on the nucleus, unless more neutrons could be given off, as envisaged above.

During pioneering experiments in neutron bombardment, many materials subjected to such bombardment changed to new isotopes or new elements. The mechanism of this process may be understood by first considering an atom of carbon, which contains six protons and six neutrons in the nucleus and six electrons circulating. If the carbon atom is bombarded by neutrons, it is converted into an isotope with six protons and *seven* neutrons, still with the six circulating electrons. This is a stable isotope, which is contained in ordinary carbon to a small extent. The specimen of carbon bombarded by the neutrons would be made to contain more atoms of 'carbon-13' than occur in natural carbon.

On the other hand, if we bombarded sodium, which has only one stable isotope and in which all the atoms contain 11 electrons, 11 protons and 12 neutrons, we should produce an unstable form of sodium with 11 protons and 13 neutrons. One neutron is surplus and cannot be held in the nucleus. It can change from the unstable condition by acquiring a positive charge which will turn it into an extra proton. Since the neutron is an electrically neutral particle it achieves this transformation by giving out a negative charge (that is an electron). This brings the atom to a condition where it now consists of 12 protons (an increase of one), 12 neutrons (same as before), and therefore the atom is no longer that of sodium, but has been converted to the next element, magnesium, which has 12 electrons.

THE USE OF URANIUM FOR NUCLEAR ENERGY

Realizing that neutron bombardment could convert an element into one with a higher atomic number, the Italian physicist Enrico Fermi had the idea of creating new elements by bombarding uranium, which, with 92 protons, was the largest known atom. The metal is composed principally of uranium-238, but less than one per cent of its mass derives from the isotope uranium-235. Following the principle described above for converting sodium into magnesium, Fermi succeeded in proving that he had converted part of the uranium, first into 'neptunium' with 93 protons, and this in turn into 'plutonium', with 94 protons; both these are synthetic elements.

In 1938 the German physicists Otto Hahn and Fritz Strassmann were studying the bombardment of uranium by neutrons. While repeating Fermi's experiments, they discovered that other elements had been formed, including the metal barium and an inert gas, probably krypton, both in such minute amounts that their existence could barely be identified. This was unexpected and at variance with the effect discovered by Fermi, described in the previous paragraph. At first Hahn and Strassmann thought they had produced only 'transuranic elements', like plutonium, which have a greater atomic number than uranium. On repeating their experiments they confirmed, but could not explain, the fact that minute amounts of lighter elements had been produced. An Austrian scientist Dr Lise Meitner, with her nephew Otto Frisch, interpreted the result as indicating that the nucleus of the uranium-235 isotope had been split into several fragments with atomic weight ranging from about one-third to about two-thirds that of uranium, coupled with the production of several secondary neutrons. In contrast with this, Fermi's experiment had changed an already large atom into one slightly larger. Hahn and Strassmann's momentous findings, as interpreted by Lise Meitner, showed that they had achieved 'fission'. Later it was realized that, among the atoms of uranium, those of the 235 isotope were behaving in a different manner from the major part of the uranium, which is principally the 238 isotope.

The weight of the nucleus of uranium-235 is about one-thousandth part heavier than the sum of the weights of the fission fragments. According to Einstein's equation the mass destroyed in the process would be converted into a great amount of energy. Only a few days before Hitler invaded Holland, Belgium, and France, this discovery was reported in the news-papers; immediately afterwards, for reasons which are now well apparent, most further references to the fission of uranium-235 were put under the seal of security until after Hiroshima, over five years later.

If the formation of the two main fragments were the only result of fission of the uranium atom, the process would be of no greater value in the provision of large-scale energy than Rutherford's early experiments. But, apart from the fragments, the fission of uranium-235 also releases two or three neutrons. Thus we reach the position envisaged on page 299 in which the fission of one atom provides two or more projectiles well suited to split more atoms, thus starting a chain reaction capable of leading to continuous self-activation, with the release of energy.

THE ATOMIC BOMB

The amount of uranium-235 contained in the small sample of uranium bombarded by Hahn and Strassmann was minute, the neutrons given out were fast-moving, and practically all of them escaped from the tiny portion of uranium before a chain reaction could take place. If, however, a mass of uranium-235 could be created, having a diameter of about 80 mm (a sphere of this size would weigh about 10 kg) this would exceed what is known as the 'critical mass'; sufficient neutrons would be given enough chance to collide with more and more uranium-235 atoms without escaping. Under these conditions a chain reaction could take place, thus annihilating a fraction of the mass and instantaneously evolving a vast amount of energy.

The construction of the atomic bomb involved a concentration of scientific and industrial effort on a staggering scale and any simplification of the problem does less than justice to the effort involved. However, simplifying as we must, the process required is to make two portions of uranium-235, each of which exceeds half the critical mass. The two portions must be separated until the required moment for explosion, when they are rammed together at a very high speed. Immediately they touch, the critical mass is exceeded and the chain reaction initiated by the neutrons takes place.

The separation of uranium-235 is an operation of prodigious complexity and cost. The chemical properties of the isotopes of an element are identical, so they have to be separated by physical methods, relying on the slightly different atomic weights. One method is to extend the use of the mass spectrograph, referred to on page 296. Another method is to make a gaseous compound uranium hexafluoride. Those molecules of the gas which contain the 235 isotope are a little lighter than the major part, containing uranium-238 and, in the gaseous form, move about or diffuse at a slightly greater rate than the heavier ones, the difference being about one per cent.

The gaseous compound is made to diffuse through very fine pores in a series of multiple diaphragms. The faster moving molecules pass through somewhat more readily than the heavier ones, so as the gas diffuses through each diaphragm the proportion of the lighter molecules is slightly increased. The process repeated over and over again gradually enriches the proportion of uranium hexafluoride containing the 235 isotope. The scale on which this process is necessary can be indicated by remembering that the plant at Oak Ridge, Tennessee, U.S.A., employed 75,000 people in a number of immense buildings, spread over seventy square miles, with a maximum output of only 3 kg of uranium-235 per day, and consumed a vast amount of electric power in the process.

Uranium hexafluoride is a highly corrosive gas and will destroy many of the normal materials of which chemical apparatus is made. The compound is itself liable to decomposition, particularly if moisture is present, and, unless stringent precautions are taken, will change to a solid material deposited on the walls of the diaphragm membranes, evolving the poisonous fluorine and hydrofluoric acid. Like some other abominable things, it is given a friendly nickname – 'hex'.

THE GAS CENTRIFUGE

Until 1968 the gaseous diffusion plants described above were the principal method of enriching uranium in the 235 isotope but in that year Britain, West Germany and the Netherlands began a joint project to develop the gas centrifuge process which will occupy less space and be cheaper to run than 'Hex' plants like Oak Ridge (U.S.A.) or Capenhurst in Britain. The centrifuge scores over the diffusion process in running cost, because the amount of electricity, though still formidable, is less than one fifth that required in gaseous diffusion of uranium hexafluoride.

The centrifuge consists of a vacuum tank containing a drum about a metre long with a central nozzle at one end and concentric orifices at the other. When the centrifuge is rotated rapidly, the heavier atoms tend to move to the circumference and the lighter ones to near the inner orifice. The enrichment is only minute but it is fed to the next centrifuge, which again causes a slightly higher proportion of uranium hexafluoride of 235 isotope to be moved away from the heavier 238 isotope.

One of the many problems of the process arises because the rotating speeds of the drum must be 60,000 rpm, several times as fast as the revolutions of the wheels of the car which broke the land speed record. Only the highest strength stainless steels or titanium alloys or perhaps carbon fibre material are strong enough but, to make the process economic, it will be essential to avoid too great a material cost.

This process shows great promise and is being actively developed throughout Europe. It has the advantage that fewer stages are required to enrich uranium and the plants will require much less power compared to the diffusion process. It is estimated that a factory using centrifuges would cost £50–£100 million to construct. However it is expected that the world's nuclear fuel industry will grow rapidly, from a turnover of about £100 million at present to £750 million in 1985.

CONTROLLED FISSION FOR NUCLEAR ENERGY

When scientists began to turn their attention from the costly production of atomic bombs to the possibilities of harnessing nuclear power for peaceful purposes, it was realized that, for economic reasons, natural uranium, or the metal only slightly enriched in the 235 isotope, must be used.

Heat is created by the fission of the 235 isotope which forms 0·7 per cent of the mass of natural uranium. This can be made to split if hit by slow neutrons, which divides the 235 isotope into two fragments, releasing heat energy and two or three fast-moving neutrons. These must be slowed down by what is called a moderator. Some of the slow neutrons hit and split other uranium atoms, thus bringing about a chain reaction, which however, unlike the conditions in the atomic bomb, is a controlled chain reaction, regulated by materials which can absorb neutrons. The effect of neutrons on the bulk of the uranium – the 238 isotope – is to transform some uranium atoms into plutonium which, like uranium, contains isotopes which, under certain conditions, are fissile.

It is worth reminding ourselves that in the atomic bomb, which was made of over 90 per cent uranium-235, fast neutrons initiated the fission. So long as the mass of U-235 was less than critical, the fast neutrons would escape before having a chance to cause fission. When the critical mass was exceeded the fast neutrons became so thoroughly trapped that the chain reaction immediately took place. Fast neutrons travel at speeds of many thousands of kilometres per second. In the controlled conditions of the nuclear reactor slow neutrons are harnessed, having speeds of only a few kilometres per second. Such neutrons are commonly called thermal neutrons.

NUCLEAR POWER STATIONS

In 1945 the British Government decided that the nuclear research centres at Harwell and at Risley should investigate the use of nuclear energy for power. By 1950 the possibilities of producing electricity in this way had been examined, and appeared likely to be economic. Three years later construction of the Calder Hall nuclear power station was begun; in the following year, 1954, the U.K. Atomic Energy Authority was set up. On 17 October 1956 Calder Hall, the world's first full-sized nuclear power station, was opened. This station was the first of several, all operating on similar principles. They are known as gas-cooled reactors. *Fig. 55* shows the general lay-out of the station. The information given below is specific to Calder Hall, in order to give some idea of the scale of these reactors, and to indicate the function of the metals that are used, but it will be realized that in any series of nuclear reactors there will be developments, improvements and changes in size.

The station consists of four reactors housed in cylindrical steel pressure vessels over 20 metres high. Each reactor shell surrounds a large graphite construction weighing 650 tonnes, made up in the form of 58,000 bricks with channels running through it, containing about 10,000 rods of natural uranium (that is, with only 0·7 per cent of the 235 isotope). Each rod is about a metre long and 30 mm in diameter; the total weight of uranium is about 120 tonnes in each reactor.

The uranium is sealed in a metal 'can' which is necessary in order to contain the waste products of fission, including the plutonium which is 'bred' during the operation of the reactor. The can is finned, to assist the transfer of heat, and it seals the uranium in order to prevent it from being oxidized by the carbon dioxide gas which is used to transfer heat from the reactor to the heat exchangers. The mundane word 'can' hardly does justice to an object which is difficult and expensive to make, and which possesses a kind of beauty, as will be seen from Plate 21a. In the cans of such gas-cooled reactors a magnesium-rich alloy is used, known as Magnox, which is discussed later. Plate 21b shows the loading of fuel elements into the machine which places them into

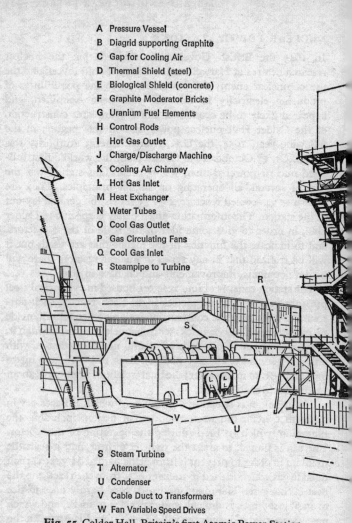

A Pressure Vessel
B Diagrid supporting Graphite
C Gap for Cooling Air
D Thermal Shield (steel)
E Biological Shield (concrete)
F Graphite Moderator Bricks
G Uranium Fuel Elements
H Control Rods
I Hot Gas Outlet
J Charge/Discharge Machine
K Cooling Air Chimney
L Hot Gas Inlet
M Heat Exchanger
N Water Tubes
O Cool Gas Outlet
P Gas Circulating Fans
Q Cool Gas Inlet
R Steampipe to Turbine

S Steam Turbine
T Alternator
U Condenser
V Cable Duct to Transformers
W Fan Variable Speed Drives

Fig. 55. Calder Hall, Britain's first Atomic Power Station

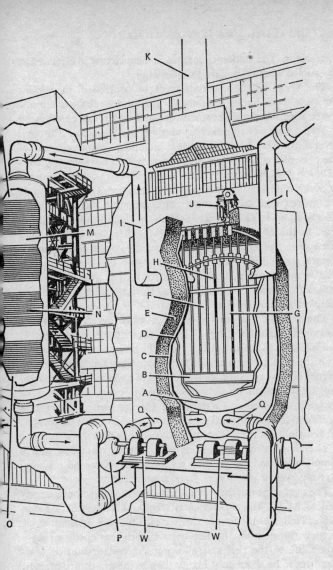

the reactor. The station concerned is Chapelcross in Scotland, a gas-cooled reactor similar to Calder Hall.

The speed of the neutrons must be controlled, to prevent excessive heat being developed; this is done by selecting a material which can entrap neutrons and which can be adjusted to keep a steady 'neutron cross-section' in the reactor. Although several materials have been used, including cadmium, most of the present nuclear power stations in Britain have control rods of boron steel, which are continuously, though slightly, adjusted in position during the operation of the reactor. The control rods are moved with a very small travel into the reactor core by electric motors. The rods are suspended electro-magnetically so that, in the event of electrical failure, they would drop fully home, their full length of 7 to 10 metres then being entirely in the reactor core; this would immediately stop the chain reaction. The normal movement of the control rods is very finely controlled to fractions of a millimetre. In the Calder Hall reactor there are 48 control rods.

The heat of the reactor is taken away by forcing carbon dioxide through the system. The gas enters the reactor at 140°C and is heated to about 340°C. One of the many facts that surprise a visitor is that as much as a ton of carbon dioxide per second circulates. The reactor core is enclosed in a welded steel pressure vessel surrounded by a concrete shield; over 10,000 tons of concrete was used at Calder Hall. Having heated the carbon dioxide in the reactor, and passed the hot gas to a heat exchanger, the remainder of the power station is conventional.

The programme of building this type of reactor is nearly complete and the eleven such stations, comprising 26 reactors, have a power production of 5,300 megawatts.

The next stage was the Advanced Gas-cooled Reactor (abbreviated to AGR); the experimental station at Windscale, near to Calder Hall, was the first of this series. This uses uranium oxide as fuel, the uranium having been slightly enriched, giving a longer life to the fuel and allowing the carbon dioxide temperatures to be about 200° higher than in the gas-cooled reactor. As Magnox is not suitable to endure this greater heat, the cans are of stainless steel. The fuel is made of uranium oxide pellets inserted into the stainless steel cans; each fuel element contains 36 cans and the reactor holds a total of nearly 30,000 cans. The use of uranium oxide was an important development; all subsequent types of reactors and, so far as is known, all those being developed no longer use fissile elements in the metallic form.

The steel pressure vessel of the experimental reactor at Windscale is contained in a spherical steel outer shell; this is partly as a safety measure, partly to facilitate experimental work. Following encouraging results from the Windscale AGR, similar designs, but in steel-lined prestressed concrete pressure vessels, were started at Dungeness, Hinkley Point, Hunterston, Hartlepool and Heysham. There have been problems and delays, but the stations at Hinkley Point 'B' and Hunterston 'B' are now producing power. The first commercial AGRs have proved costly to build; there have been unexpected problems with corrosion of some boiler materials, with vibration and noise inside the reactor, requiring costly thermal insulation and special methods of inspecting inaccessible components. However the AGR in a prestressed concrete vessel is safe, efficient and easy to operate and there are good prospects that satisfactory output will be achieved.

Fast reactors use a different concept in which no moderator is used. The first experimental reactor was established at Dounreay and has been producing about 60 megawatts since 1959. The fast reactor fuel is a mixture of oxides of uranium and plutonium and, under the conditions existing in the fast reactor, some of the isotopes of plutonium which would not fission in a thermal system will do so in the fast reactor. The considerable heat evolved is removed by molten sodium which has a high conductivity and which, in spite of some of its rather unpleasant characteristics, has proved an excellent material for rapidly extracting heat from the fast reactor. A pilot station called the Prototype fast reactor has been constructed at Dounreay and has started putting power into the grid system. It will pave the way for the first commercial fast reactor, which will be ordered when experience has been gained.

Table 22 shows some details of these three types of reactors.

Table 22

	Gas-Cooled Reactor	Advanced Gas-Cooled Reactor	Fast Reactor
TYPICAL SITE	Calder Hall	Windscale	Dounreay
NUCLEAR FUEL	Natural uranium	Enriched uranium oxide	Mixed uranium and plutonium oxides
MODERATOR	Graphite	Graphite	None
FUEL CANS	Magnox	Stainless steel	Stainless steel
CONTROL RODS	Boron steel	Boron steel	Tantalum or boron carbide (under investigation)
COOLANT	Carbon dioxide	Carbon dioxide	Liquid sodium
SIZE OF REACTOR (cubic metres)	2,340	580	6

The comparison in size of reactor core is interesting – the gas-cooled reactor the size of a medium sized cinema, the advanced reactor a small bungalow and the fast reactor only the size of a dustbin. The fast reactor produces forty times more electricity per ton of uranium than gas-cooled reactors. It is also much more efficient in the use of our uranium resources. The gas-cooled reactors utilize a maximum of two per cent of the uranium which is put into the reactor; fast reactors could, theoretically use all the available uranium, though it must be noted that the uranium from which uranium oxide is formed is enriched and therefore more costly.

OTHER REACTORS

There are many other types of nuclear reactors, including prototypes for later developments, university and industrial research reactors. The latter are often designated by christian names based on the initials of the system that is being used; for example Horace is the H_2O Reactor Aldermaston Critical Experiment and Vera the Versatile Experimental Reactor Assembly.

In Britain the Steam Generating Heavy Water Reactor is being developed. The fuel elements are of pellets of slightly enriched uranium oxide canned in zirconium alloy. In the place of graphite, heavy water (page 297) is used as moderator and, in the place of carbon dioxide, ordinary water is used to convey the heat from the reactor to the electricity generators. Like the gas-cooled reactors, the SGHWR turns part of the uranium into plutonium and is therefore a valuable working method of breeding this synthetic element which will be required in great quantity, in the coming generation, for fast reactors. The first SGHWR was built at Winfrith and started production in 1968.

The experimental Dragon Reactor, funded by ten nations, was built at Winfrith to investigate the principles of high temperature gas-cooled reactors. The core of the reactor consists of 37 fuel elements each consisting of seven rods joined in a cluster. The reactor is designed so that a number of different fissile materials can be used, including thorium compounds. The fuel is in the form of particles of uranium oxide or other experimental materials; a complex coating is added to each particle, comprising carbon with a silicon carbide interlayer. As the coating is ceramic there are no stringent limitations to the operating temperature, which is normally 1,250°C and can run up to 2,000°C for short periods. The coolant gas is helium.

In the U.S.A. conventional electric power generation is cheaper than in Britain. When nuclear power stations were first developed, in the mid-1950s, it was a natural step for the Americans to develop the system already employed as a submarine-propulsion unit – the pressurized water reactor (PWR). Enriched uranium dioxide sealed in tubes of zirconium alloy, is contained in a massive steel pressure vessel 3 metres in diameter and 12 metres high. The fuel is 'cooled' by water at 300°C at a very high pressure and this superheated water is used in boilers to produce steam. A variant of this type, the boiling water reactor (BWR) produces steam directly in the reactor vessel for use in the power unit; in this case the steam is temporarily radioactive. Westinghouse, and U.S. General Electric have had orders from thirty countries, and over two hundred of these types of reactors are now built or on order. There remains some controversy about long-term safety but the American reactors seem likely to produce electricity more cheaply than the AGR and with a lower capital cost.

THE ROLE OF THE METALLURGIST IN NUCLEAR ENERGY

Metallurgically uranium has a number of peculiarities. It oxidizes rapidly in air at 200°C and above; its atomic lattice structure at room temperature is complex and each crystal of the metal expands non-uniformly when heated. This leads to internal stresses when the metal undergoes heating and cooling.

Two distortion effects can occur in uranium owing to the process of fission. 12–15 per cent of the fission product consists of the inert gases xenon and krypton, which tend to collect in fine bubbles, and exert internal pressures which can lead to large volume changes. Another distortion problem connected with the properties of uranium is growth due to the effect of radiation. One direction in the uranium crystal gets longer while another contracts. This appears at its worst in rolled uranium and can cause the bar to double its length. The effect can be minimized by arranging that the grains of uranium are very small and in random orientation. As mentioned above, the modern reactors use oxides of uranium or other fissile elements and this eliminates metallurgical problems due to growth.

CANNING MATERIALS

The choice of the correct material for the cans required in a nuclear power station depends principally on the type of reactor and especially on the working temperature inside the reactor, but a number of other technical problems are involved. The metal must not capture too many neutrons otherwise the efficiency of the reactor will be lowered. The cans must be able to withstand the cooling-gas environment at the operating temperature. The canning material should conduct heat reasonably well and must withstand the temperature at which the reaction is controlled, up to 800°C in advanced gas-cooled reactors. It must be capable of being formed into thin-walled containers, with fins to aid dissipation of heat, and it must be sealed effectively after the uranium fuel has been inserted.

The canning material must be sufficiently ductile to allow for deformation and distortion in the reactor. It must have enough strength to support its own weight; it must not warp under the effect of heat or radiation or under the stresses caused by warping of the uranium or the external load of other fuel elements. The can must contain no porosity or other discontinuities. There is continuous monitoring of the solidity of the cans in the reactor so any leak can be quickly detected and the offending can removed. Therefore a leak is not dangerous, though a definite nuisance, because it would reduce efficiency. Once an atomic reactor has been started, by-products, including intensely radioactive isotopes of strontium and barium, and gases krypton and xenon, are formed. Therefore the cans must be completely reliable while they are immured in the reactor.

When the uranium is 'spent', the contents of the can must be capable of being dissolved in chemical agents, by remote control, so that the uranium can be reclaimed for further use and plutonium recovered. The cans themselves have to be buried in a silo because, by the time their job in the reactor is finished, they are highly radioactive, and likely to remain so for many hundreds of years.

Magnesium was an ideal material for the first design of gas-cooled reactors, as it complied to a greater or lesser extent with the stringent requirements listed above. 'Magnox', a magnesium alloy, containing about 0·8 per cent aluminium and 0·01 per cent beryllium, is used to sheathe the uranium fuel in those reactors. The small additions to the magnesium ensure that a more tenacious and protective oxide skin is formed than would be obtained with pure magnesium. Several other 'Magnox' alloys have been developed, with different aluminium contents;

one such alloy contains 0·7 per cent manganese, and another contains 0·55 per cent zirconium.

As the tendency develops to processes which evolve greater heat, differing canning materials are required. Magnox would not be suitable at temperatures higher than those existing in the first design of gas-cooled reactor. Until recently it was thought that beryllium might be an ideal material for cans in the advanced gas-cooled reactor, but metallurgical problems, as yet unsolved, caused the large-scale production of beryllium to be suspended. Stainless steel has the advantages of resisting the heat and not warping; it absorbs more neutrons than beryllium and has a lower heat-conductivity. To get optimum results, a stainless steel can has to be only about a third of a millimetre in wall section thickness. The advanced gas-cooled reactor at Windscale uses stainless steel, in the shapes of tubes about 15 mm diameter, with a number of fine flutings.

Zirconium is a useful material for nuclear applications. It absorbs less neutrons than stainless steel, and has good corrosion-resistance in water reactor environments. This metal has what is known as good 'compatibility' with uranium oxide, a problem that is referred to later.

SOME MISCELLANEOUS PROBLEMS

The control rods have to be mechanically strong and must have a very high absorption of neutrons. In the gas-cooled reactors the control rods are of stainless steel tubes lined with a 3 per cent boron steel sleeve and a 4 per cent boron steel core. At Calder Hall there are 48 rods in the reactor, each weighing 60 kg. For more compact or more sophisticated reactors other materials, sometimes exceedingly costly ones, are chosen for their greater ability to absorb neutrons. Hafnium has recently become more available, and is preferred for compact reactors. By far the best absorber of neutrons is the metal gadolinium (one of the 'rare earths', see page 254) but supplies are so scarce that its use is limited to special reactors which require control rods of extra-ordinarily high efficiency.

To withstand the internal pressure and to ensure that no harmful radiation products escape, the welding of the steel sheet, about 100 mm thick, and the subsequent heat-treatment to relieve internal stresses, are precise and important operations. Special attention is given to what is called 'compatibility' which involves similar problems to sacrificial corrosion, mentioned on page 271. The metals in the nuclear power unit must not deteriorate through contact with any other materials. For example, the stainless steel structures in a fast reactor will 'sit' in hot sodium for thirty years and be exposed to fast neutrons for the same period, all in addition to the thermal stresses which they must endure.

As in all industry there is a continual succession of small, medium and large problems. Like a blast furnace or a rolling mill the nuclear power station has to be kept running with a maximum of economy, output and safety. Breakdowns are annoying and costly and every change in operating conditions may introduce the possibility of some problem arising. For example, a corrosion problem which has become rather important in reactors using Magnox is the oxidation of nuts and bolts. Those which are within the primary circuit are made of ordinary steel which was completely satisfactory in all the earlier stations. However, developments of the Magnox systems increased the gas temperature, so that the oxidation rate of the steel became too high. The immediate solution was to down-rate the newer gas-cooled reactors so that the gas temperature was lowered by about 10°C. However, even after being down-rated the Magnox stations are proving to be the cheapest method of producing electricity. These are only a few of the problems, nearly all of them unsuspected twenty years ago, which confront the metallurgist in the nuclear energy industry.

The wasting sources of coal, oil and gas provide an incentive to the use of nuclear power, but the economic factors must not be neglected, and conventional power stations will have a spur to encourage greater efficiency and lower cost. Under conditions of inflation any cost figure may soon become out of date but, as a general statement, a gas-cooled reactor station can produce power as cheaply as a conventional one; an AGR more cheaply. Probably fast reactor stations will be even more competitive. It is intended that with the present programme, by 1977 Britain's nuclear power stations will produce 11 million kilowatts, representing about a fifth of our electric power requirements.

22

THE FUTURE
OF METALS

We are thinking of the year 2000, over fifty years since the first edition of this book appeared. Possibly man's metallurgical endeavours may be dedicated to bigger and worse ways of self-destruction. However it is more likely that common sense will prevail, so we will look at the future with some optimism. In order to discuss what kinds of metals will be needed in the future, it may help to remind ourselves of man's basic needs – to sustain life, to make life comfortable, to have a measure of privacy but to be able to communicate in many ways with other people, to work, to enjoy leisure, to travel, to create and see beautiful things.

Assuming that we have scope to fulfil these needs, it is likely that more metals will be required for vehicles on land, on sea and in the air; for an increasing variety of domestic appliances; for all the girders, pipes, tubes, cables, wires and sheets that are required to make homes and places where people work or otherwise congregate; for untold quantities of the mesh and rods which strengthen reinforced concrete and for the beautiful shapes of metal that make bridges.

It is expected that by A.D. 2000, Britain's requirements for electric power will be four or five times that of 1970. This will mean more power stations, cable, dynamos, motors, switchgear, instruments, factory buildings and nuts and bolts. As depressed parts of the world population raise their standard of living, they will increase their use of energy – and therefore materials – at an even faster rate than the so-called civilized countries.

Such further advances will need industrial complexes strategically placed across the world in places where the basic energy sources are used and metals and materials are extracted.

New steel-works are likely to be capable of producing 10 to 20 million tons of steel products per year from one site, adjacent to a deep sea harbour. Electricity, whether from thermal energy or nuclear energy, is likely to be generated in 1,000 megawatt units of 3 to 5 per site. Such massive installations will need medium-sized efficient works to finish, fabricate and assemble the metals and materials in a multitude of appliances.

Most families will have at least two vehicles; the petrol-driven one will be for long distances; the other will be electric-battery powered, for shopping and short trips and to enable commuters to get to the bus or rail stations. Both types will need batteries and it is expected that interesting developments will take place in this industry. In the last five years the electric power per unit weight in the lead–acid battery has nearly doubled. It only needs a further 50 per cent improvement in efficiency to make econ-omic an electric car or locomotive capable of a hundred-mile trip. As an example of the progress that has been made, a prototype electric bus has been running from Birmingham to Manchester, a distance of over 90 miles.

In competition with these batteries, exotic combinations of elements are being investigated. The zinc–air battery seems the most likely possibility, but the sodium–sulphur battery has been tried in many parts of the world. When one realizes that both the sodium and sulphur are molten in this type of battery, the prospect seems almost crazy. However, the last few years have seen liquid sodium being used as a coolant in fast reactors so it may soon be commonplace – though such materials must be kept meticulously out of harm's way.

Before the end of the twentieth century we shall probably have six or eight weeks' annual holiday, so there will be time to travel to interesting parts of the world by ship, instead of squashing inside jet aircraft. If we get bored by hotels, we shall need camping equipment, caravans, safari and rough country vehicles, all requiring metal tubes, sheets and castings. Some people may become even lazier and rely on bigger and better television sets to justify holidays at home; such TV might be equipped with the 'feelies' described by Aldous Huxley (plenty of transistors required!) or the two-way service so ominously envisaged by George Orwell. The sets will become smaller, but the screens will get bigger.

The wider scope of education in schools, technical colleges and universities and the greater opportunities for leisure activities will bring out new talents in the community. In any case a civilized person creates his own reaction to some of the more monotonous aspects of modern times. He commutes all the week, but buys a dinghy for weekend sailing; he does a repetition job in a mass production industry and reacts by spending spare time in model making, home decoration or making artificial jewellery. This will undoubtedly cause increased interest in what were regarded as old-fashioned metals and some very satisfying crafts such as wrought iron work, copper, pewter and other sheet-metal beating and perhaps home kits for investment casting.

COMPETITORS OF METALS

Why does man use metals – or timber or concrete? Apart from aesthetic considerations a material is used because it offers the required strength and other properties at minimum cost. In working out the strength factors we need to know the cost of the material per kilogram, the density of the material, and its strength, which is usually stated as a force per unit of cross-section. But most materials are not purchased in terms of cross-section; more usually one buys metal by the tonne and concrete by the load.

Over the last two hundred years users of materials have been subconsciously driven to select a combination of those properties which are cheapest for the specific application. For most large engineering structures strength is the most important and expensive item. It is not surprising, therefore, that because reinforced concrete is the cheapest way of buying strength, it is far and away the major tonnage and volume material used by mankind. Of course, reinforced concrete depends on its strengthening, hidden core of steel. Second is timber, which is a remarkably cheap way of buying strength. The next best bargain is steel, and structural steel in particular, followed by aluminium and plastics, in that order.

Since plastic materials are not very competitive on a cost-per-unit strength basis it is strange that so much has been said about an all-plastics world. Perhaps the modern methods of salesmanship and public relations have helped this but it is time that plastics are seen in their proper perspective, at least in engineering usage. Plastics are manufactured in vast quantities and hitherto have cheapened as production increased, compared with metals, which have been increasing in cost. Crude oil distillates from which plastics are made are now less plentiful and more costly, which has led to plastic price increases and shortages.

Another serious handicap concerns the problem of what to do with all the plastic scrap. A plastic article may have a useful life of only a few weeks, in the case of packages, or of a few years in the case of components. Injection moulded articles can be shredded and recycled, provided that the materials are not mixed up. Compression moulded plastics are difficult to recycle and plastic sheets are a positive embarrassment. Recycling is the Achilles' heel of the plastics industry; it is a tremendous problem when one considers the vast volume. The great bulk of plastics have to be buried, burned or dumped in the sea. There is a possibility that some process may be developed for shredding used plastic and making it into chipboard. Another possibility is coating chipped plastics with asphalt and using it for hard fill, perhaps for road building. We must comment that these limitations are stressed because the plastics industry says very little on the subject; over-selling of plastics always has been a rather boring aspect of an otherwise interesting industry.

It is increasingly evident that metals and plastics can often be combined. For corrosion-resistance, steel pipes lined with plastics are being used on an increasing scale and plastic-coated steel is used for building. Structural partitioning is made by bonding wood dust with plastic, and coating it on both sides with aluminium strip. But the combination of metals with plastic brings a problem of reclamation more difficult than the recycling of plastics.

SOURCES OF METALS

Helped by the developments in transport and earth-moving, remote parts of the world such as central Africa, western Australia, and Brazil are being opened up to win new and rich sources of metals. The techniques which are evident in motor-way construction have shown how large masses of earth can be dug out of one place and moved elsewhere, at the same time land-scaping the surrounding environment attractively. In New Zealand, where they remove iron-ore sand from beaches of North Island and turn it into a slurry, the areas which have been depleted are afterwards re-landscaped by contract.

The winning of metal ores will be affected by new and exciting technical developments. Open-cast mining, less hazardous than underground working, will become increasingly productive as techniques of earth-moving improve. Sea water is already an inexhaustible source of magnesium, but the riches on the sea-bed are only just beginning to be explored. Mineral nodules, like big pebbles, containing 5 to 30 per cent manganese, with appreciable amounts of copper, nickel, cobalt and iron, have been found, in concentrations of 15,000 tonnes per square kilometre, at depths of over 3,000 metres. Preparations are being made to dredge areas between southern California, Nicaragua, Panama and Hawaii. But does the sea bed belong to one company, or nation, or to all mankind?

The collection, sorting and conversion of metal articles into reprocessed ingots is already a vast industry in steel, cast iron, aluminium and copper. Previously the source of such material was from the rag and bone man's collection of old saucepans, bedsteads, broken down lawn mowers and cheap tin trays. This involved a great deal of sorting before a pure metal or alloy could be produced. More sophisticated plants for sorting, crushing or shredding scrap metals and transporting them economically to the refining furnaces will be developed, though a great deal of reliance will still be placed on those experienced people who have the knack of distinguishing one alloy from another of similar appearance.

The process developed in Belgium for separating steel from

non-ferrous metals in crushed scrap motor-cars shows the kind of ingenuity that is being used to deal with scrap. The old car is crushed to a rectangular block, which is then showered with liquid nitrogen. At this very low temperature, non-ferrous metals remain ductile, but iron and steel become as brittle as glass, and break into small splinters which can then be separated by electromagnets. Mechanical processes for shredding scrap motor-cars are being developed in many countries, where magnetic and other processes separate the ferrous and the non-ferrous parts.

Scrap of any metal in any country ought to be preserved and recycled with the same attention and control as if it were the metal in virgin condition. Yet often this is neglected. An illustration of this neglect can be quoted from the copper industry. During the past 35 years over 50,000 tons per year of scrap copper and copper alloys have been shipped from Britain to Europe for refining and making into castings and wrought metals. Much of this copper was brought into Britain, probably during the first two decades of the twentieth century, at prices near to £100 per ton. This copper, having been utilized for a generation in electric motors, dynamos, or copper alloy castings, may now be sold at £500 or more per tonne as scrap, which might at first sight appear to be very good business. But since, during the next few years, copper may increase in price to over £1,000 per tonne, the selling of copper scrap abroad may be short-sighted.

During the next generation, the major metal industries of the world – steel, aluminium and copper – will recycle more than 50 per cent scrap in their manufacturing processes. The increasing costs of accurately sorting metal scrap may lead engineers and metallurgists to discuss the widening of specifications. There are cases where the accuracy of composition of an alloy must be controlled to very fine limits, but in many other alloys too close a specification is pedantic. Alloys of aluminium and of copper, for example, can tolerate certain impurities in relatively large amounts without undue change in mechanical or other properties.

THE PRICES OF METALS

The values of metals are influenced by their comparative availability in the earth's crust, the cost of their production, and the scale of their industry. *Table 23* shows the prices for raw metals, provided by the well-known journal, *Metal Bulletin*. Some prices – for copper, tin and lead for example – are those quoted on the London Metal Exchange (LME). Some – like aluminium, nickel and magnesium – are producer prices in industries where one producer is historically sufficiently dominant to set the market for the others. Zinc is an interesting case of a metal whose representative price once came from the LME, but in recent years a producer price has been quoted by the *Metal Bulletin* after study of individual selling price announcements by leading producers. LME prices are quoted daily and tend to fluctuate quite widely in the short term. Producer prices move less often, but by bigger steps when they do change; both fundamentally respond to the law of supply and demand. In recent years currency instability has been a prime cause of apparent instability in metal prices.

REFRACTORIES AND CERAMICS

Since the earliest days of metallurgy, refractories have played a vital part in metallurgical operations, because without them it would have been impossible to use the high temperatures required for the efficient extraction and refining of metals. Whereas, in the past, refractories have been considered as ancillary to metal production, they are now taking on a new interest and becoming important 'end products' combined with metals. Requirements of jet turbine blades to operate at high temperature are so exacting that it may be found, in the not too distant future, that a mixture of refractory metal oxide bonded with a metal may exhibit better service behaviour than a heat-resisting alloy.

The ceramics industry has always been aligned closely to metallurgy; their problems are somewhat akin and the one is dependent upon the other. Most high-temperature metallurgical

Table 23. Average Metal Prices £ Sterling per Tonne

Year	Pig Iron	Semi-finished Steel	Aluminium	Copper	Lead	Tin	Zinc	Magnesium	Nickel
1939	5	8	94	43	15	226	14	201	180
1945	7	12	89	62	28	302	29	201	187
1950	10	17	114	179	106	745	119	201	354
1955	16	25	167	352	106	740	91	266	511
1965	22	32	196	469	115	1,413	113	229	632
1970	26	38	257	588	126	1,530	123 (128)*	340	1,211
1971	31	45	257	444	104	1,438	127 (150)*	363	1,246
1972	33	47	230	428	103	1,506	151 (173)*	356	1,430
1973	35	57	273	726	174	1,960	345 (215)*	480	1,414
1974	47	70	379	876	252	3,494	527 (332)*	919	1,797
1975	65	89	420	557	185	3,091	335 (390)*	919	2,415
1976	85	110	579	782	250	4,255	395 (468)*	1,339	3,286

Copper, lead, tin, and zinc are given as LME yearly averages. Ferrous prices are year's average of U.K. market. Others are given as producer price at year-end
* Zinc prices in brackets are GOB producer prices at year-end.

operations, such as refining, melting, pouring, and casting, must be carried out in containers lined with refractory ceramic materials. In recent years, a closer tie between ceramics and metals has developed because of the need for materials which are capable of carrying stresses at much higher temperatures than has been necessary hitherto. Such applications are now required for heat engines, space missiles, nuclear reactors, and electronic components.

Alloys for High-Temperature Service

Molybdenum, tungsten, tantalum and niobium alloys will be used in appropriate temperature ranges, sometimes alone, sometimes coated with a thin layer of ceramic. Some estimates of the temperature ranges at which various alloys can be used are given below.

The use of metals at high temperature will be facilitated by

Table 24. Probable Future Temperature Ranges of Materials

Material	Temperature °C
Tungsten-base and tantalum-base alloys	1,500–2,000
Molybdenum-base alloys	1,200–1,400
Niobium-base alloys	1,100–1,350
Chromium-base alloys	to 1,200 but limited to very low stresses above 1,000
Complex alloys containing two or more of the above metals as a base	1,650
Nickel-base and cobalt-base alloys	1,000
Iron-base alloys	900
Titanium-base alloys	to 650 for times up to 10 minutes or less
Beryllium-base alloys	to 600 for special applications where light weight and high elastic modulus are important

cooling with gases or liquids. Another way of taking full advantage of the properties of materials with high strength at elevated temperatures will be by making honeycomb structures of thin metal walls. This involves developing methods for making ultra-thin strips and progressively welding, or high-temperature brazing them, so that abutting walls are joined into a honeycomb form. This procedure is already exploited for 'skinning' the fuselages of aircraft and sophisticated space vehicles, because of the very great strength and rigidity obtainable, coupled with a high capacity to absorb vibration.

Already the demand for materials capable of withstanding high temperatures has resulted in the development of a new family of materials which contain a metal combined with carbon, nitrogen, silicon, or boron. Small particles or fibres of these refractory compounds are held together with a ductile metal acting as a cement. Being a combination of ceramic and metal, these materials are known as 'cermets' and are formed into blocks, closely approaching the shape required for their end use.

METALLURGY AS A CAREER

A generation ago metallurgists were more or less confined to their laboratories, so opportunities for advancement were not plentiful. The change has been gradual and is still continuing, but now metallurgy as a career opens up prospects of management and of being in charge of processes, instead of being an advisory boffin. The opportunities in the future will lie in five main fields.

Mining and Production

All over the world labour costs are increasing and the cost of energy is becoming greater. In 1973 there was a belated realization that existing cheap sources of energy were limited and new sources will take years to develop. Many prices of metals and materials increased substantially and some temporarily became scarce. These higher prices led to the exploitation of new deposits and a reappraisal of weak ores, for example those of tin in Cornwall, which previously were too difficult to win economically. At the same time the increased values of some by-products have been having a substantial effect on the profitability of extraction processes; for example small percentages of silver and mercury, which are obtained during zinc smelting operations, help these processes, which were not very profitable, to become more viable.

While these developments have been affecting the economics of metal smelting, problems of contaminating the environment have become of major concern. Fume emissions, working conditions and good relationships with local inhabitants will henceforth affect the future of any new or existing metal smelter. This can only be done at great expense, which may lead to uneconomic manufacture. The recent substantial rise in zinc prices has been caused by the closing of five major American units, because of contamination of the environment. The efforts to produce metal economically have been linked with new methods of discovering ore deposits, excavation, transport and concentration of ores and new methods of smelting the ores with less energy per tonne than before. There is no doubt that such endeavours, coupled with a more efficient re-cycling of scrap metals, will become one of the biggest challenges for the metallurgist of the future.

The Metallurgist in the Consumer Industries

In this field, firms which design, engineer or construct will employ the metallurgist or materials technologist (it is a pity that, so far, nobody has thought of an acceptable single word for this). These 'materiologists' will be specialists in the melting, alloying, shaping, treatment and service behaviour of ferrous and non-ferrous metals. They will regard other materials – concrete, rubber, stone – as requiring the same kind of study as metals. Part of their training will be in Value Analysis in which the choice of materials, their dimensions and assembly, are scrutinized from first principles.

The nuclear power metallurgist can be included in this group, though the metals that once were classified as exotic are widely used in nuclear engineering and need a study of their own. Furthermore the paramount needs of safety bring problems that were never encountered before. When eventually atomic fusion, as distinct from fission, is harnessed for creating power, a new series of apparently insuperable problems will arise.

Management

The opportunities for metallurgists in management are already developing. This is a great change for the better; not so long ago, the metallurgist, though qualified and necessary, was out on a limb and did not get many opportunities to progress directly into management. This was partly caused by the training of metallurgists a generation ago. The authors of this book can remember being sent, when they were students, for a fortnight each year to whichever metal works would condescend to accept them as visitors. This training usually developed into helping with simple analyses in the laboratory, helping to pay the wages on Fridays or fishing in the nearby canal because no suitable work could be found.

University courses are now being designed to make the student aware of the opportunities and responsibilities in management. It is not uncommon for a student in his first or second year to work for some months in a staff position in a works, his duties being discussed by the management, the university staff and the student. When he returns to the university for the remainder of his course, the student has a much better idea of 'what it is all in aid of' and has learned at first hand the problems which he will encounter later.

The question of cost enters more and more into university studies as, of course, this is a vital part of management. Students are now required to design factory lay-outs for a given process, to estimate the cost of the building, machinery, power and materials and to link their studies of the project with visits to factories which are concerned with the same problems.

There will always be scope for the management-metallurgist in large firms but increasing opportunities are arising where the combination of technical knowledge, cost consciousness and management training can equip a young man for promotion in a smallish company. Great opportunities will exist in the future for women in metallurgy. No doubt a generation ago the heat of processes, the danger of splashing metal, the robustness of the language and the 'man's world' of metallurgy combined to make women feel out of place. However, processes are changing, metal rolling is becoming automated and computerized; mathematics, electrical engineering and other disciplines are becoming associated with metallurgical processes and the whole environment and conditions of work in these industries are more acceptable for women than in the past.

Metallurgical Research

No new natural metals remain to be discovered and it is unlikely that many inexpensive new alloys with absolutely outstanding properties will be developed. However you never can tell; for example when this book was first published the superplastic alloys had not been heard of but they will probably require several paragraphs in our next edition. A typical superplastic alloy is based on zinc containing 22 per cent aluminium. The material in sheet form is heated to about 300°C and then quenched to room temperature. When reheated to 250–275°C it develops a fine grain size and becomes superplastic with an elongation of as much as 1,000 per cent.

Research on metallurgical processes is continuous, always stimulating, though often disappointing when a new process that went well in the laboratory develops problems in full-scale production. The direct reduction of iron ores to steel, continuous processes for making sheet metal, new smelting processes and the extension of powder metallurgy to the making of large components are a few of the multitude of projects.

The state of mind needed in industrial research is what we might describe as 'the man from Mars' approach. Such a man might visit a manufacturing company, and, because he had never seen the process before and had no preconceived ideas, he might notice the illogicalities or follies of that process. Those who are concerned every day with a process are often limited by immersion in their local problems.

In research on processes there are at least three stages, each of which needs this 'man from Mars' approach. First the process is worked out on a small scale under laboratory conditions to establish the principles. Next a medium sized prototype plant is built, operating perhaps in the works but still under careful control and measurement. Finally the large capital expenditure for a full scale plant is authorized. As the size of the process increases, new problems arise, which need an objective and self-critical attitude. For example, doubling the linear dimensions of a container multiplies the volume of that container by a factor of eight, or, as we remarked when discussing large Japanese blast furnaces, doubling the inner surface area of a furnace quadruples the volume. Sometimes such increase of scale is beneficial, as with new blast furnaces. Sometimes it brings new and costly problems in its train, as in the new smelting processes for zinc. In all developments, however, changes in scale or conditions introduce problems which call for clear and unprejudiced thinking. The research and development of new metallurgical processes offer great opportunity for young and clever enthusiasts who realize that while the process is being perfected there is no substitute for personal attention, early in the morning and late at night.

The Metallurgist as Sherlock Holmes

The examination of failures or flaws, and the prevention of accidents by knowing what to look out for, is an interesting and challenging service. The development of non-destructive testing has been discussed in chapter 9 and involves many new and sophisticated techniques. The investigation of steam boilers that have exploded or bridges that have collapsed, either during construction or in service, will require very detailed micro-structural examination and testing of failed parts. Other examples are motor-car accidents, which often require forensic science, a fund of metallurgical knowledge and the exercise of common sense. Even more satisfying, however, is the prevention of failure. If the effect of fatigue in the early Comet aircraft, or the harmonic vibrations in the Tacoma Bridge, or the potential weaknesses in box girder bridge construction, could have been foreseen, a great deal of suffering and waste of money would have been prevented.

In finishing a chapter on the future and coming to the end of a book, it is a temptation to indulge in a peroration. However, one of the exciting things about science is that you never know what will happen tomorrow. Far greater scientists than we have announced what they thought was the 'last word' on their branch of knowledge and have soon been proved to be wrong. Therefore modestly but hopefully we will wait and see.

GLOSSARY

(Technical terms which are indexed and amply defined in the text are not included in the glossary.)

AGEING. As applied to castings in steel and cast iron, the word indicates a period of time provided to relieve casting stresses. Ageing is also used in reference to wrought aluminium alloys of special composition; after heat-treatment they are quenched in water and then kept at room temperature for a period of four to six days. During this period, their maximum strength and mechanical properties are fully developed.

ALLUVIAL. Deposited by rivers.

ALPHA PARTICLE. The nucleus of the helium atom, containing two neutrons and two protons, emitted from the nuclei of certain radioactive elements.

ANGSTROM UNIT. A unit of measurement used by metallurgists and crystallographers, giving the distance between atoms in a space lattice. An angstrom unit is one ten-millionth of a millimetre.

ANNEALING. The process of heating a metal or alloy to some predetermined temperature below its melting point, maintaining that temperature for a time, and then cooling slowly. Annealing generally confers softness.

BINARY ALLOY. An alloy composed of only two major ingredients, e.g., copper and zinc, or lead and antimony. An alloy with three ingredients is known as a 'ternary alloy'.

BREAKING DOWN. The first stage in the shaping of an ingot of metal, with the object of reducing its section and refining its grain structure.

BRINELL HARDNESS. A comparative measure of the hardness of metals. A hard steel ball is impressed on the smoothed surface of the metal being tested, under a given load and for a fixed time. The diameter of the impression of the ball is measured and the value referred to a chart or conversion table, which gives the Brinell hardness of the metal. The figure ranges from about 10, in the case of soft metals like lead, up to 600 and higher for heat-treated steels.

CARBIDE. The compound formed when an element combines with carbon. The carbides of metals are usually intensely hard.

CATALYST. A substance which, when present in small amounts during a chemical reaction, promotes the reaction, but itself remains unchanged.

CEMENTITE. The name given to identify one constituent in iron–carbon alloys. Cementite is essentially iron carbide, Fe_3C, but may contain other substances such as manganese and chromium, carbides of which are dissolved in the iron carbide and do not appear separately.

CERMETS. Materials produced by bonding a ceramic, such as an oxide, carbide, nitride, or boride, with a metal or alloy. The bonding is effected at high temperature under controlled conditions, by methods similar to those used in powder metallurgy.

COMPATIBILITY. A term used especially in nuclear engineering. Two materials are said to be compatible when they can exist in contact with each other without interaction.

CONDUCTIVITY (ELECTRICAL or THERMAL). The measure of the ability of a substance to allow the passage of electricity or heat. Copper is an example of a good conductor, rubber of a bad one.

CORES. Specially fashioned pieces of sand or metal used to form the hollow parts of a casting. To make a cylindrical hole in a casting, a cylindrical solid core of similar shape is used.

DEEP DRAWING. A shaping process in which the whole or part of a disc of metal is forced through the aperture of a die so as to make a cup shape. By repeating this process with plungers and apertures of progressively decreasing diameters, the metal is drawn into an elongated cup or closed tube.

DIELECTRIC. A non-conductor of electricity.

DIES. Metallic or other permanent forms which confer a given shape on a piece of metal. The word covers a range of meaning and includes dies which are used to shape solid metal in presses and those for making diecastings from liquid metal.

DISTILLATION. The conversion of the whole or part of a liquid substance into gaseous form, to be followed by the subsequent condensation of this to liquid.

DOLOMITE. A widely occurring whitish mineral; its chemical composition is magnesium carbonate – calcium carbonate.

DUCTILITY. The property of a metal which enables it to be given a considerable amount of mechanical deformation (especially stretching) without cracking.

ELECTRODE. A conductor which conveys electric current directly into the body of an electric furnace, plating vat, or other electrical apparatus.

ELECTROLYSIS. A process involving chemical change, caused by the passage of an electric current through a fluid solution.

ELECTRON. Elementary negatively charged particle having a mass about 1/1,840 that of a hydrogen atom.

ELECTRON MICROSCOPE. This is a form of microscope which is capable of very high magnification. For the detection of the finest details of a structure, light, the exploring medium of an optical microscope, has too coarse a 'texture' (wavelength). The electron microscope employs an exploring medium of a much finer 'texture' – a stream of electrons. The stream can be rendered parallel or convergent by 'lenses' consisting of magnetic fields. The electron stream is scattered by the object to be examined and an image is formed on a fluorescent screen or on a photographic plate. Magnifications of up to 100,000 can be obtained.

ELEMENTS. All compounds can be resolved into elements; thus, water into hydrogen and oxygen; common salt into sodium and chlorine. The elements, however, cannot be resolved by chemical means into any simpler substances. Including the modern 'synthetic elements' which have been made in the atomic physicist's laboratory there are over a hundred elements, of which more than three-quarters are metals.

EQUILIBRIUM. The state of balance which exists or which tends to be attained after a chemical or physical change has taken place. Equilibrium may not be reached for long periods after the change has been initiated.

FLASH. When a metal is forged or cast in a die some metal penetrates the space where two die surfaces touch and thus a web of metal, known as 'flash', remains attached to the forging or casting and has to be removed by filing or clipping.

FLUX. A chemical, used to combine with a substance having a high melting point, generally an oxide, forming a new compound which can readily be melted.

GAMMA RADIATION. Rays of very short wavelength, less than that of X-rays, produced during the disintegration of radioactive materials.

HIGH-FREQUENCY ELECTRIC FURNACES. Furnaces in which the metal to be heated is held in a container which is surrounded by an electric circuit, through which alternating current of high frequency passes. This high frequency current 'induces' currents in the metal charge, which cause it to melt by internal resistance. The frequencies used are of the order of 1,000 to 1,000,000 cycles per second, compared with the frequency of 50 used in domestic heating and lighting in most parts of Britain.

HOT SHORTNESS. An undesirable property of certain metals and alloys whereby they are brittle in some elevated temperature range.

INCLUSION. A non-metallic particle of slag, oxide, or other chemical compound which has become entangled in metal during its manufacture.

INGOTS. Blocks of metal made by casting the liquid metallic contents of a furnace or crucible into open metallic moulds.

ISOTOPES. Atoms of the same element, having the same number of electrons and protons but having different numbers of neutrons. The isotopes of an element are identical in their chemical and physical properties except those determined by the total mass of their atoms.

MACH NUMBER. The relation between the speed of a moving body to the local speed of sound, which can vary with temperature, altitude, and therefore pressure. The terms are generally used in connection with the speeds of supersonic aircraft. Mach 1 is the speed of sound at sea level; the speed of sound is reduced with increasing height. The word is derived from the name of Ernest Mach, a 19th-century mathematician and philosopher.

MAGNETITE. A black, magnetic iron ore. Deposits of a very high grade magnetite occur in Sweden.

MALLEABILITY. A property of metals enabling them to be hammered and beaten into forms such as that of thin sheets, without cracking. Gold is the most malleable of all metals.

MEGAWATT. One million watts.

METALLOGRAPHY. The study, observation, and photographing of the structure of prepared specimens of metals, usually with the aid of a microscope. From such a study, much can be learnt about the condition, heat-treatment, and manufacturing history of metals.

METALLOID or 'SEMI-METAL'. An element which has some properties characteristic of metals, others of non-metals. Examples are arsenic and boron.

METALLURGY. The art and science of producing metals and their alloying, fabricating, and heat-treatment. The word can be pronounced 'metall*urgy*' or 'met*a*llurgy', though the former may be preferred in keeping with the adjective 'metall*u*rgical'.

MICRON. A millionth part of a metre: a thousandth part of a milli-metre.

MICROSTRUCTURE. The structure of a metal seen under the micro-scope.

MODULUS OF ELASTICITY. The ratio of stress to strain in a material. The strain is usually a measure of change of length. The stress is a measure of the force applied to cause the strain.

NEUTRON. Electrically uncharged particle possessing a slightly greater mass than the proton. Neutrons are constituents of all atomic nuclei except that of the normal hydrogen atom.

OXIDE. A chemical compound formed when an element unites with oxygen, as by the action of burning. Water is hydrogen oxide; sand, silicon oxide; and quicklime, calcium oxide.

PIG IRON. Crude iron as produced from the blast furnace and con-taining carbon, silicon, and other impurities. On flowing from the furnace the molten iron is run into channels which branch out into a number of offshoots about $1300 \times 130 \times 130$ millimetres. The iron runs into these and takes the familiar form of 'pigs'. The main channels are called 'sows'.

PROTON. Positively-charged particle in the nucleus of an atom having a mass about 1,840 times that of the electron and an electric charge equal in magnitude to the negative charge of the electron.

PYROMETERS. Instruments for measuring high temperatures.

REAGENT. A substance which is added to another in order to bring about and take part in chemical action.

REFRACTORIES. Firebricks or other heat-resisting materials used for lining furnaces and retaining the heat without allowing the outer shell of the furnace to be damaged. Refractories are grouped into 'acid', 'base', and 'neutral' according to their composition and their action on the hot substances with which they come into contact in the furnace. Examples of the three types are silica, dolomite, and carborundum respectively.

RETORT. A specially shaped vessel intended for containing substances which are to be heated to form a vapour which is then collected or condensed.

ROASTING. The process of heating an ore at medium temperature in contact with air.

SEMI-CONDUCTORS. Materials which at room temperature have much lower electrical conductivities than metals, but whose conductivities increase substantially with increase of temperature. This is in contrast with the electrical behaviour of metals, whose electrical conductivities decrease slightly with increase of temperature. Germanium, selenium and silicon are semi-conductors.

SLAGS. Glass-like compounds of comparatively low melting point, formed during smelting when earthy matter contained in an ore is acted on by a flux. If the earthy matter were not deliberately converted into slag it would clog the furnace with unmelted lumps. The fusibility and comparatively low density of the slag provides a means by which it may be separated from the liquid metal.

SMELTING. The operation by which a metallic ore is changed into metal by the use of heat and chemical energy.

SOLUTION. The intermingling of one substance with another in so intimate a manner that they are dispersed uniformly among each other, and cannot be separated by mechanical means. Although solid substances dissolved in water are the most familiar solutions, the word has a wider meaning; for example, gases dissolved in liquids, liquids dissolved in liquids, and, frequently in metallurgy, solids dissolved in solids.

SPECIFIC GRAVITY. The ratio of the density of a substance to the density of water.

SPECTRUM. The result obtained when radiation is resolved into its constituent wavelengths. The spectrum of white light, consisting of coloured bands, is the most familiar, but all types of radiation comprising a range of wavelengths can be resolved into spectra.

STRAIN. A measure of the amount of deformation produced in a substance when it is stressed.

STRESS. A measure of the intensity of load applied to a material. Stress is expressed as the load divided by the cross-sectional area over which it is applied.

TAPPING. The controlled removal of liquid metal or slag from a furnace, the metal running through a tap-hole near the base of the furnace.

TEMPERING. A warming process intended to alter the hardness of a metal which has already been subjected to heat-treatment. The tempering temperature is lower than that at which the first heat-treatment was carried out.

THERMIT. A mixture of powdered iron oxide and aluminium which, when ignited, sets up a vigorous chemical action, whereby the aluminium unites with oxygen from the iron oxide, leaving metallic iron. A high temperature develops, so great that the iron is melted.

THERMOCOUPLE. When two wires of different metals are joined at each end and one junction is heated, a small electric voltage is produced which depends on the temperature and which can be measured by a delicate instrument. Certain combinations of metals are specially suitable, by virtue of the amount of current produced and their resistance to heat. Thermocouples are used for some types of pyrometers.

THERMO-PLASTIC. A plastic material which can be melted or softened repeatedly without change or properties. Injection mouldings are made with thermo-plastic materials.

THERMO-SETTING. A plastic material which can be softened by the combination of heat and pressure and then be compression-moulded into a required shape. Once having been treated in this way the thermo-setting material does not revert to its former condition and cannot be re-treated.

TROY. System of weights used for gold and silver. A kilogram contains 32·15 troy ounces.

X-RAYS. Radiation of a character similar to that of light, but of a shorter wavelength and possessing the property that it can pass through opaque bodies.

ZONE REFINING. A method of purifying metals, usually in small quantity. A bar of the solid metal is progressively melted by intense local heating; during subsequent solidification some of the impurities diffuse towards the portion last to solidify. By repeating this operation several times the metal becomes progressively purified, the impurities being segregated at one end of the bar. Metals can be zone-refined to contain less than one part in a million of impurity. Germanium has been zone-refined to one part of impurity in a hundred thousand million.

INDEX

More about Penguins and Pelicans

Penguinews, which appears every month, contains details of all the new books issued by Penguins as they are published. From time to time it is supplemented by *Penguins in Print*, which is our complete list of almost 5,000 titles.

A specimen copy of *Penguinews* will be sent to you free on request. Please write to Dept EP, Penguin Books Ltd, Harmondsworth, Middlesex, for your copy.

In the U.S.A.: For a complete list of books available from Penguins in the United States write to Dept CS, Penguin Books, 625 Madison Avenue, New York, New York 10022.

In Canada: For a complete list of books available from Penguins in Canada write to Penguin Books Canada Ltd, 2801 John Street, Markham, Ontario, Canada L3R 1B4.

Mathematics in Western Culture

Morris Kline

'The good Christian should beware of mathematicians,' admonished St Augustine; and his distrust is echoed today by all those who are allergic to figures.

If you have never agreed with Plato that arithmetic 'has a very great and elevating effect' on the soul, you may doubt that mathematics is anything but a series of dry technical procedures. But you could be in for a surprise, because in this lively history of the cultural influence of mathematics Professor Morris Kline shows how, without mathematics, there would be no art, music, philosophy or religion as we know them. His informed discussion ranges from Ancient Egypt, where priests used their knowledge of the calendar to predict when the Nile would flood, to Renaissance Italy, where such artists as Uccello and Leonardo da Vinci were the most skilled mathematicians of their day, and to the twentieth century, when Einstein's mathematical work on relativity changed the whole course of science.

'He is unfalteringly clear in explaining mathematical ideas ... an exciting and provocative book' – *Scientific American*

The Penguin Dictionary of Physics

Edited by Valerie H. Pitt

The Penguin Dictionary of Physics is an abridgement, specially designed for a Penguin readership, of the latest thoroughly revised version of Gray & Isaacs' *A New Dictionary of Physics*. It includes information gleaned from the latest research, and it is intended to provide a concise accurate guide to the terminology of Physics and related disciplines. Longer entries on words of major importance go beyond mere definition to discuss their significance in relation to other words and ideas within the wider context of science. SI units are used throughout.

Asimov's Guide to Science

In Two Volumes

Asimov's Guide to Science is an encyclopedic survey by one of the foremost science writers of our time, which when first published was greeted as 'the most exciting and the most readable account . . . of modern science'. Isaac Asimov's outstanding success as a writer of science-fiction sufficiently suggests his gifts for exciting narrative and simple exposition of abstruse facts.

This Pelican edition offers an invaluable work – as much for reading as for reference – in two volumes. In the first, Dr Asimov deals clearly and enthusiastically with the history of astronomy and our present understanding of earth, space and stars; the nature of matter and the atom; chemistry and the elements; and the gradual application of this knowledge in technology.

In the second volume he explains the progress and present state of the biological sciences, dealing with the mysteries of living cells and the chemistry of life; evolution, behaviour and the human brain.

Books published by Penguin
on Science and Technology